Atmospheres

Atmospheres

Richard M. Goody

Harvard University

James C. G. Walker

Yale University

Prentice-Hall, Inc., Englewood Cliffs, New Jersey

Illustrations by Richard Kassouf

PRENTICE-HALL INTERNATIONAL, INC., London
PRENTICE-HALL OF AUSTRALIA, PTY. LTD., Sydney
PRENTICE-HALL OF CANADA, LTD., Toronto
PRENTICE-HALL OF INDIA PRIVATE LIMITED, New Delhi
PRENTICE-HALL OF JAPAN, INC., Tokyo

10 9 8 7 6 5 4 3 2

FOUNDATIONS OF EARTH SCIENCE SERIES

A. Lee McAlester, Editor

ISBN:

0–13–050088–7 (p)

0–13–050096–8 (c)

Foundations

of Earth Science Series

Elementary Earth Science textbooks have too long reflected mere traditions in teaching rather than the triumphs and uncertainties of present-day science. In geology, the time-honored textbook emphasis on geomorphic processes and descriptive stratigraphy, a pattern begun by James Dwight Dana over a century ago, is increasingly anachronistic in an age of shifting research frontiers and disappearing boundaries between long-established disciplines. At the same time, the extraordinary expansions in exploration of the oceans, atmosphere, and interplanetary space within the past decade have made obsolete the unnatural separation of the "solid Earth" science of geology from the "fluid Earth" sciences of oceanography, meteorology, and planetary astronomy, and have emphasized the need for authoritative introductory textbooks in these vigorous subjects.

Stemming from the conviction that beginning students deserve to share in the excitement of modern research, the *Foundations of Earth Science Series* has been planned to provide brief, readable, up-to-date introductions to all aspects of modern Earth science. Each volume has been written by an authority on the subject covered, thus insuring a first-hand treatment seldom found in introductory textbooks. Four of the volumes—*Structure of the Earth, Earth Materials, The Surface of the Earth,* and *Earth Resources*—cover topics traditionally taught in physical geology courses. Four more volumes—*Geologic Time, Ancient Environments, The History of the Earth's Crust,* and *The History of Life*—treat historical topics. The remaining volumes—*Oceans, Man and the Ocean, Atmospheres, Weather,* and *The Solar System*—deal with the "fluid Earth" sciences of oceanography and atmospheric and planetary sciences. Each volume, however, is complete in itself and can be combined with other volumes in any sequence, thus allowing the teacher great flexibility in course arrangement. In addition, these compact and inexpensive volumes can be used individually to supplement and enrich other introductory textbooks.

Contents

4

Winds of global scale   71

*Atmospheres in Motion. General Circulation of the Terrestrial Atmo-
sphere. Weather and Turbulence. Pressure Forces. A Direct Thermal
Circulation. Circulation in the Atmosphere of Venus. The Equatorial
Hadley Cell. The Coriolis Force. A Balanced Wind. Circulation of the
Martian Atmosphere. Planetary Waves.
Retrospect.*

5

Condensation and clouds   106

*Clouds and the Radiation Balance. Saturated Vapor Pressure. Relative
Humidity, Supersaturation, and Supercooling. The Ascent of Moist Air.
Condensation on Other Planets. Condensation Nuclei. Growth of Cloud
Droplets. Rain from Warm Clouds. Snow, Hail, and Rain
from Freezing Clouds.*

6

The evolution of atmospheres   126

*The Cycles of Carbon Dioxide. Carbon Dioxide on Venus. The Run-
away Greenhouse Effect. Escape of Atmospheric Gases to Space. Terrestrial Life and Atmospheric Oxygen.*

Preface

There is a great difference between research in the laboratory and studies of the Earth and planets. In the laboratory the scientist can perform controlled experiments, each carefully designed to answer questions of his own choosing. Except in minor respects, however, the Earth and planets are too large for controlled experimentation. All we can do is observe what happens naturally and attempt to interpret this behavior in terms of the laws of physics and chemistry.

For this reason, the study of the planets is a valuable adjunct to the study of the Earth. We are denied the opportunity to perform controlled experiments in Earth science, but we can at least observe how planets not very different from Earth behave in different circumstances. By comparing observations of the different planets, we are led to new ideas about the manner in which atmospheric motions, for example, depend on the rate of rotation of the planet, on the distance from the Sun, and on the composition of the atmosphere.

In the last decade there have been great strides in our knowledge and understanding of the atmospheres of the planets, stimulated by the flights of instrumented space probes to the nearby planets and by the development of improved instrumentation for ground-based astronomy. It is now for the first time possible to give a general account of why planetary atmospheres behave as they do. We have tried to present such an account in this book. We have divided the subject matter into chemical properties, temperature, motions, cloud formation, and evolution, devoting a chapter to each area. In each chapter we discuss the physical and chemical processes that influence the atmospheric property under consideration and illustrate the discussion with specific examples drawn from our knowledge of the atmospheres of the different planets. We believe this approach provides insights into atmospheric phenomena that might be missed if we were to restrict our discussion to one particular atmosphere. Nevertheless, our emphasis is on the terrestrial atmosphere, because it is the only atmosphere that can be observed in detail. We have placed little emphasis on the weather, however, because terrestrial meteorology has been treated in a number of good books on both the elementary and advanced level; our readers will profit by collateral reading of one of these texts.

Our approach to the subject is deductive rather than descriptive. By this we mean that we have attempted to show how the properties and behavior of planetary atmospheres may be deduced by means of general arguments based on the laws of physics and chemistry. Except in the first chapter, where we have gathered together the basic data on which the rest of our discussion is based, we have held the description of atmospheric phenomena to a minimum. In principle, it should be possible to derive from first principles all aspects of atmospheric behavior, knowing only the composition and mass of the atmosphere, the size, the surface properties, the rate of rotation of the planet, and the distance of the planet from the Sun. We are not yet able, and may never be able, to carry this program to completion; atmospheres are altogether too complicated. But it is our belief that the attempt to do so is both interesting and instructive.

Our approach may seem inconsistent with the fact that the atmospheric sciences are, for the most part, applied sciences related rather directly to questions of human needs. We believe that this aspect of the subject should be emphasized, even if it is not directly reflected in our exposition. We have found in elementary classes that the fundamentals, as developed in this book, form a satisfactory basis for excursions into such matters as photochemical smog, contamination by the SST, rainmaking and climate control, and some weather phenomena.

1

The sun and the planets

In this chapter we shall consider what we need to know about a planet and its atmosphere before we can begin to discuss atmospheric behavior. It is of prime importance to know what is in the atmosphere, that is, the chemical composition. The composition of a planetary atmosphere depends on the way in which the planet and its atmosphere were originally formed, as well as on physical and chemical processes that continually add some gases to the atmosphere while removing others. Examples of these processes are volcanic eruptions, which release gases to the atmosphere, chemical reactions at the ground, and the escape of gases from the top of the atmosphere into space.

After we have discussed the processes that govern atmospheric behavior, we shall be better able to consider how the atmospheres of the planets originated and how the compositions of these atmospheres have evolved during the life of the solar system. For this reason we shall defer to Chapter 6 discussion of how the atmospheres of the planets developed their different compositions. For the time being we limit ourselves to a description of what is known about these compositions.

In order to specify the overall composition of a planetary atmosphere, we have to decide what to include. On Venus and Mars this decision is simple because there is a sharp transition at the surface of the planet from gas above to solid below. It is clear that the surface marks the bottom of the atmosphere. On the other hand, there are no clearly defined bottoms to the atmospheres of the *outer planets*—Jupiter, Saturn, Uranus, and Neptune. These planets are composed almost entirely of hydrogen and helium, as is the Sun. The gas is probably converted to a liquid or solid form in the deep interiors of the planets where pressures are very high, but the transitions are likely to be gradual. So when we talk of the atmospheres of these planets we mean the outer regions, with no attempt being made to distinguish between the atmosphere and the rest of the planet.

On Earth, the picture is complicated by the presence of the oceans. It is not always sufficient to say that the bottom of the atmosphere is the surface of the sea or land because there is a continual exchange of matter, energy, and momentum between the sea and the air above it. This exchange makes the behavior of each one dependent, to some extent, on the behavior of the other. For some purposes, therefore, it is convenient to consider ocean and atmosphere as parts of the same physical system.

Atmospheric Compositions

The composition of terrestrial air has been measured in detail, at least near the ground, and we present a summary of the more important constituents in Table 1–1. These data are representative of the overall composition of the Earth's atmosphere, although variations in composition occur at high altitudes. Nitrogen accounts for approximately 80% of the volume, with oxygen constituting most of the remainder. Many of the other gases do not significantly affect the behavior of the atmosphere, and they will be ignored by us. There are traces of water vapor, ozone, and carbon dioxide that are important, however, because these gases absorb infrared radiation emitted by the ground. Water vapor, ozone, and carbon dioxide, therefore, have an effect on atmospheric temperatures that we shall discuss in Chapter 3.

As Table 1–1 indicates, the water vapor content of the atmosphere is variable. Water vapor condenses readily to water at the temperatures that prevail on Earth, so the concentration of water vapor in air depends on the temperature and on the proximity of such bodies of water as the ocean. If the average surface temperature on Earth were higher than it is, the oceans would evaporate and the composition of the atmosphere would be changed

Table 1–1

Composition of the Earth's Atmosphere

Constituent	Per Cent by Volume or by Numbers of Molecules of Dry Air
Nitrogen (N_2)	78.084
Oxygen (O_2)	20.946
Argon (A)	0.934
Carbon dioxide (CO_2)	0.031
Neon (Ne)	1.82×10^{-3}
Helium (He)	5.24×10^{-4}
Methane (CH_4)	1.5×10^{-4}
Krypton (Kr)	1.14×10^{-4}
Hydrogen (H_2)	5×10^{-5}
Nitrous oxide (N_2O)	3×10^{-5}
Xenon (Xe)	8.7×10^{-6}
Carbon monoxide (CO)	10^{-5}
Ozone (O_3)	up to 10^{-5}
[Water (average)	up to 1]

drastically by the addition of a large quantity of water vapor. For this reason we must not overlook the ocean when we consider the composition of the atmosphere.

Let us now turn to the compositions of the atmospheres of the other planets. Our information about the outer planets is far from complete because they have not yet been visited by scientific spacecraft, and we must therefore rely on data provided by telescopic observation of the spectrum of sunlight reflected by the planets.

What data we have are presented in Table 1–2. Hydrogen (H_2), methane (CH_4), and ammonia (NH_3) have all been detected spectroscopically, but the presence of helium (He) is inferred indirectly and remains to be proven. Helium does not absorb visible light and so cannot be directly detected with ground-based telescopes.

Information for Mars and Venus is much more definite. Their atmospheres are composed largely of carbon dioxide, with minor components listed in Table 1–3. These data have been obtained by ground-based spectroscopic measurements for Mars and for the minor constituents on Venus. For carbon dioxide on Venus, however, we have direct chemical measurements made by the Venera spacecraft.

Table 1–2

Composition of the Atmospheres of the Outer Planets

(The number of molecules above the clouds
in a vertical column with a cross sectional area of one square centimeter)

	H_2	He	CH_4	NH_3
Jupiter	1.8×10^{26}	$<9.1 \times 10^{25}$	1.2×10^{23}	2.6×10^{22}
Saturn	3.7×10^{26}	—*	9.4×10^{23}	$<6.7 \times 10^{21}$
Uranus	1.3×10^{27}	—	9.4×10^{24}	—
Neptune	—	—	1.6×10^{25}	—

*Dashes indicate that no measurement is available, not that the particular gas is absent.

The Barometric Law

We have discussed the relative proportions of the different gases present in the atmospheres of the planets, but this information by itself does not provide a complete description of what is in each atmosphere. We must also consider how much gas each planet has around it; that is, we must discuss the masses of the different atmospheres.

In order to evaluate the mass of a planetary atmosphere it is necessary to understand the relationship between *atmospheric density* (the mass of gas per unit volume) and *atmospheric pressure* (the force per unit area experienced by a surface exposed to the gas). It is also necessary to understand how atmospheric density and pressure vary with altitude. The variation of pressure and density with altitude is described by the *barometric law*.

Let us compare the atmosphere and the ocean, noting some similarities

Table 1–3

Composition of the Atmospheres of the Inner Planets

(per cent by volume or by numbers of molecules)

	CO_2	N_2,A	O_2	H_2O	HCl	HF	CO
Venus	95	<5	$<4 \times 10^{-3}$	10^{-2}	10^{-4}	2×10^{-6}	2×10^{-2}
Mars	>50	<50	10^{-1}	up to 10^{-1}*	—†	—	10^{-1}
Earth	.03	79	21	up to 1	—	—	10^{-5}

*The abundance of water vapor in the Martian atmosphere varies markedly with the season.
†Dashes indicate that the gas is not present in detectable amounts.

and one very important difference. Although both are fluids* and both are bound to the Earth by the force of gravity, the ocean has a finite depth. We can refer to the surface of the ocean, but we cannot refer to the "top" of the atmosphere because there is no upper surface to the atmosphere. Instead, the atmosphere blends slowly into interplanetary space.

The reason for this difference between the ocean and the atmosphere is that the atmosphere is compressible but the ocean is not. By compressible we mean that atmospheric gases expand and contract as pressure varies. Thus the density of the atmosphere varies with the pressure, whereas the density of ocean water hardly varies at all.

The way in which the density of a gas varies as the pressure is varied has been established by experiment in the laboratory. The results of such experiments may be summarized by a single statement known as *Boyle's law*: gas density is almost exactly proportional to gas pressure. Another set of experiments is summarized by *Charles' law*: if the pressure is held constant, gas density is inversely proportional to temperature. These two descriptions of the behavior of gases can be combined in a relationship called the *ideal gas law*: density (gm cm^{-3}) is proportional to pressure (dyne cm^{-2}) divided by temperature (deg K),

$$\text{Density} = \frac{\text{Pressure}}{\text{R} \times \text{Temperature}} \tag{1-1}$$

In this equation R (erg gm^{-1} deg^{-1}) is a constant of proportionality called the *gas constant*. For sea water the equivalent expression for the density is

$$\text{Density} = \text{Constant} \tag{1-2}$$

Now let us see how an ocean and an atmosphere differ because of the different behavior of their densities. To do so we will perform an imaginary experiment with a barometer. Let us lower our barometer to the bottom of the sea. The pressure that it measures depends on the depth of the ocean and the density of the water. The weight of the water in the ocean exerts a force on the sea bed. This force (dyne) is given by the mass (gm) of the water times the acceleration due to gravity (cm sec^{-2}; according to *Newton's laws of motion*, force equals mass times acceleration). The mass of water is the density (gm cm^{-3}) multiplied by the volume, which is the product of depth (cm) and surface area (cm^2). Since pressure is force per unit area (dyne cm^{-2}), we can also write the total force as pressure times the surface area. If we equate these two expressions for the total force, we find *Pressure × Area = Density × Depth × Area × Gravitational acceleration* or

$$\text{Pressure} = \text{Density} \times \text{Depth} \times \text{Gravitational acceleration} \tag{1-3}$$

*A solid is rigid; adjacent parts of a solid cannot move with respect to one another. Relative motion of the different parts of a fluid is possible, on the other hand; this is what is meant by *flow*.

Having made this measurement, let us raise our barometer until it reads a pressure that is exactly one-half the pressure recorded at the bottom of the sea. According to Eq. (1–3), we have to halve the depth of the barometer in order to do so. Now let us raise the barometer far enough to halve the pressure again. This time the barometer need only be raised by one-quarter of the ocean depth rather than by the one-half that the first maneuver required. We continue in this fashion, halving the pressure by halving the depth of the barometer. Each time the distance the barometer moves is smaller (see Fig. 1–1).

For the second half of our experiment we will work in the atmosphere instead of in the ocean. First let us measure the pressure of the atmosphere at the ground; then we raise the barometer to the level in the atmosphere where the pressure is exactly one-half the pressure at the ground. When we raise the barometer far enough to halve the pressure again, we find a result very different from that of the ocean experiment. We have to raise the barometer as much the second time as the first time. In order to halve the pressure a second time, the height of the barometer has to be doubled. This is a major difference between atmospheres and oceans. We can repeat this process, raising the barometer far enough to halve the pressure, as often as we choose, and each time the height of the barometer will increase by approximately the same amount (see Fig. 1–2).

In order to see why this is so, consider how pressure and density change as height increases. Between the ground and the first level, pressure decreases by an amount equal to one-half the pressure at the ground. As in the case of the ocean, this pressure decrease is the gravitational force per

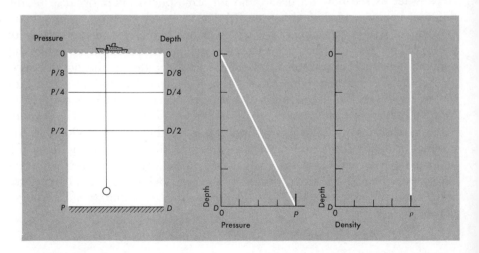

FIGURE 1-1 (At left) An experiment to measure the depth of various pressure levels. (At right) The pressure and density as a function of depth in the ocean.

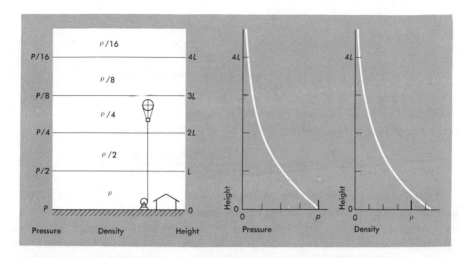

FIGURE 1-2 *(At left) An experiment to measure the height of various pressure levels. (At right) The pressure and density as a function of height in the atmosphere.*

unit area on a vertical column of material between the ground and the height of the first level. We can express this in the form of an equation equivalent to Eq. (1–3)

$$1/2 \times \text{Ground pressure} = \text{Density} \times \text{Height increase} \times \text{Gravitational acceleration} \tag{1–4}$$

where density refers to the average density of the gas between the ground and the first level. For the next upwards step we have to reduce the pressure from one-half the pressure at the ground to one-quarter the pressure at the ground, which is a reduction by an amount equal to one-quarter the pressure at the ground. This is half the reduction of the previous step. But the average density of the atmosphere in the second layer is only one-half as great as the average density in the first layer. According to Boyle's law, density is proportional to pressure, and the pressure in the second layer is only one-half as great as the pressure in the first layer. So, for the second layer of atmosphere, the equation becomes

$$1/4 \times \text{Ground pressure} = 1/2 \times \text{Density} \times \text{Height increase} \times \text{Gravitational acceleration} \tag{1–5}$$

which gives the same value for the height increase as did Eq. (1–4).

And so it goes. For each succeeding layer, the pressure reduction on the left of the equation is halved, but the density on the right of the equation is also halved because the atmosphere is compressible. The result is that the height increase in each layer is the same.

The sun and the planets

We can rearrange Eq. (1-4) to obtain an explicit expression for the height increase.

$$\text{Height increase} = \frac{1/2 \times \text{Pressure}}{\text{Density} \times \text{Gravitational acceleration}} \qquad (1\text{-}6)$$

If we now replace density in this equation by the expression for the density given by the ideal gas law (Eq. 1-1), we find that the height increase is proportional to temperature divided by gravitational acceleration

$$\text{Height increase} = \frac{R \times \text{Temperature}}{2 \times \text{Gravitational acceleration}} \qquad (1\text{-}7)$$

As we have shown, *height increase* is the thickness of an atmospheric layer over which density and pressure decrease by a factor of 2. In atmospheric science it is the convention to deal with an equivalent quantity, the thickness of a layer over which density and pressure decrease by a factor of $e = 2.7182$. This thickness is called the *scale height*. In terms of the scale height, the barometric law may be stated as follows: Atmospheric density and pressure decrease by a factor of e (which equals a decrease by 62.2%) every time the altitude increases by one scale height.*

The expression for the scale height resembles Eq. (1-6)

$$\text{Scale height} = \frac{R \times \text{Temperature}}{\text{Gravitational acceleration}} \qquad (1\text{-}8)$$

In the form we have derived, the barometric law is strictly true only for an isothermal atmosphere, an atmosphere in which the temperature does not vary with height. However, the law is modified only slightly for an atmosphere in which temperature is different at different heights. For most purposes we may ignore the modification, provided we recalculate the scale

---

*The barometric law can be stated mathematically in terms of the exponential function,

$$\frac{p(z)}{p(0)} = e^{-z/H}$$

where $p(z)$ is the pressure at height z, $p(0)$ is the pressure at the base of the atmosphere, and H is the scale height. If we set z successively equal to $0, H, 2H, 3H$ etc., we find $p(z)/p(0)$ equal to 1, e^{-1}, e^{-2}, e^{-3}, and so on. That is, the pressure decreases by a factor of $e = 2.7182$ for each increase of height equal to H. An equivalent statement of the barometric law in terms of the height increment L that corresponds to a decrease in pressure by a factor of 2 is

$$\frac{p(z)}{p(0)} = 2^{-z/L}$$

where L is the quantity we have called *height increase* in Eqs. (1-4) to (1-7).

The discerning reader may find a discrepancy between our definitions of scale height (H) and height increase (L). If his curiosity is aroused, he should ask why our derivation of Eq. (1-7) is only approximately correct.

height at each altitude, using, in Eq. (1–8), the temperature of the atmosphere at that altitude.

Let us work out the value of the scale height near the bottom of the Earth's atmosphere. An average value for the temperature of the air near the ground is 288°K, and the acceleration due to gravity at the surface of the Earth is 981 cm sec^{-2}. The value of the gas constant must be calculated because it depends on the mass of the gas molecules. Air is a mixture of four parts nitrogen to one part oxygen and has a mean molecular mass of 29 atomic mass units (1 atomic mass unit $= 1.66 \times 10^{-24}$ gm). In C.G.S. units the gas constant is equal to 8.314×10^{7} divided by the mean molecular mass of the gas expressed in atomic mass units. For air, therefore, the gas constant is 2.87×10^{6} erg deg^{-1} gm^{-1}.

We now have values of all the numbers appearing on the right-hand side of the expression for the scale height (Eq. 1–8). Substituting these values into the equation, we calculate that the scale height of the atmosphere near the ground is equal to 8.4×10^{5} cm or 8.4 km. Thus, at a height of 8.4 km in the Earth's atmosphere, the pressure is only 38% ($= 100/2.7182$) of its value at the ground. The reduction in density is comparable.

In order to get a better feeling for the way in which pressure and density decrease with altitude in accordance with the barometric law, let us assume that the temperature, and therefore the scale height, are the same at all heights. At a height of 84 km, therefore, the pressure is reduced by a factor of e raised to the power of 10 for the 10 scale heights. This factor is 22,000. Despite this enormous decrease in pressure and density, we will find that phenomena at 84 km and even higher levels are important to an overall understanding of atmospheres.

The most convenient way to represent the variation of pressure and density with altitude is in a graph that shows the logarithm to the base 10 of the pressure or density plotted against altitude. Such a graph for the Earth's atmosphere is shown on the left in Fig. 1–3. The curves in this graph would be straight if the scale height did not change at all with altitude.* The reason the curves are not exactly straight is that the temperature of the atmosphere varies with height as shown on the right. Equation (1–8) tells us that

---

*The slope of the curve is inversely proportional to the scale height. This we can see by taking the logarithm of both sides of the barometric law (see footnote on p. 8).

$$\log \frac{p(z)}{p(0)} = \log p(z) - \log p(0)$$

and

$$\log e^{-z/H} = -\frac{z}{H} \log e$$

Therefore,

$$\log p(z) = \log p(0) - \frac{z}{H} \log e$$

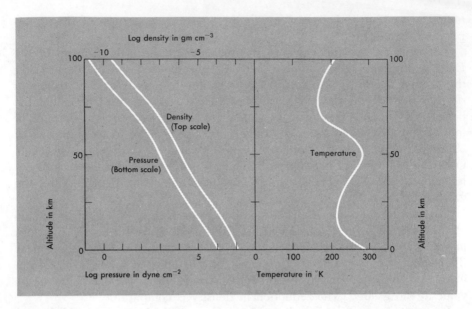

FIGURE 1-3 *The logarithm of pressure, the logarithm of density, and the temperature plotted against altitude in the Earth's atmosphere.*

the scale height is small where the temperature is low, which means that the pressure and density decrease rapidly with increasing altitude. Conversely, where the temperature and therefore the scale height are large, the curves in Fig. 1–3 drop off less steeply.

Temperature is a most important atmospheric parameter, if only because of its influence on the rate of decrease of pressure and density with altitude. In Chapter 3 we will discuss the factors that control atmospheric temperature. Equation (1–8), however, shows that the scale height also depends on the gas constant (R). Thus the scale height depends on the mean molecular mass (and therefore the composition of the atmosphere), as well as on the acceleration due to gravity (which is determined by the mass and radius of the planet). All three factors—temperature, composition, and gravitational acceleration—contribute to differences in the scale heights of the atmospheres of the different planets.

As an illustration, in Table 1–4 we present values of the quantities that determine scale height for a number of planets where the data are reasonably firm. It happens that these scale heights are all quite close to one another. On Mars the low value of gravitational acceleration is counteracted by the high molecular mass of carbon dioxide. On Venus the high molecular mass is counteracted by the high surface temperature. On Jupiter the low temperature and the high gravitational acceleration are counteracted by the

Table 1–4

Scale Heights of Planetary Atmospheres

	Gas	Mean Molecular Mass (amu)	Gravitational Acceleration (cm sec⁻²)	Average Surface Temperature (°K)	Scale Height (km)
Venus	CO_2	44	888	700	14.9
Earth	N_2, O_2	29	981	288	8.4
Mars	CO_2	44	373	210	10.6
Jupiter	H_2*	2*	2620	160†	25.3

*There may be enough He (4 amu) on Jupiter to increase the mean molecular mass significantly and hence to decrease the scale height.
†The temperature near the top of Jupiter's clouds.

low molecular mass of hydrogen. Because the scale heights are not very different, curves for the *inner planets* (Venus, Earth, and Mars) showing the logarithm of density as a function of altitude are roughly parallel (see Fig. 1–4).

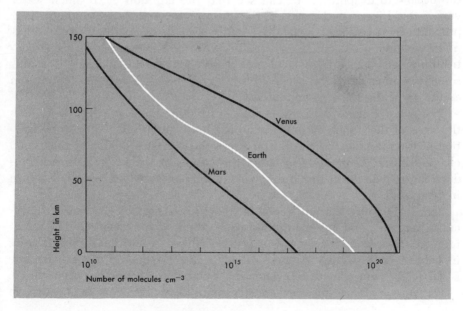

FIGURE 1-4   *Altitude profiles of the number of gas molecules per cubic centimeter in the atmospheres of Venus, Earth, and Mars.*

The sun and the planets

Atmospheric Masses

Now that we know how the gas in a planetary atmosphere is distributed with height above the surface, we can again take up the question of how much gas the different planets have in their atmospheres. We need to know the pressure at the bottom of the atmosphere because, as we pointed out in our discussion of the barometric law, the pressure is equal to the weight per unit area of the overlying material. We can, therefore, calculate the mass of the atmosphere above unit area of surface by dividing the pressure by the acceleration due to gravity. Multiplication by the total area of the surface gives the total mass of the atmosphere.

The average sea level pressure on Earth is 1.013×10^6 dyne cm^{-2}*. Dividing by the acceleration due to gravity (981 cm sec^{-2}) gives 1.03×10^3 gm as the mass of atmosphere above one square centimeter of surface. The surface area of a sphere is 4π times the square of the radius, so the Earth, which has a radius of 6371×10^5 cm, has a surface area of 5.10×10^{18} cm^2. The total mass of the Earth's atmosphere, therefore, is 5.29×10^{21} gm†. The mass of the oceans, for comparison, is 1.35×10^{24} gm.

Suppose now that the surface temperature were high enough to cause the oceans to evaporate. We can see that the addition of so much water vapor would increase the mass of the atmosphere, and therefore the surface pressure, by a factor of more than 250. Under these conditions the atmosphere would be composed principally of water vapor, with nitrogen contributing less than one molecule for every 500 water molecules. For this reason we must keep the ocean in mind when we discuss the composition and mass of the Earth's atmosphere.

The mass of the solid Earth is 5.98×10^{27} gm, much greater, even, than the mass of the ocean. Clearly the atmosphere comprises a very small fraction of the total mass of the Earth. From these masses it follows that small changes in the composition of either the oceans or the solid Earth can produce large changes in the atmosphere, a subject to which we shall return in Chapter 6.

The surface pressure on Venus has been measured by the Soviet spacecraft, Venera 7, and has been found to be about 80 atmospheres. The gravitational acceleration on Venus is given in Table 1–4, and the radius appears in Table 1–5. With these data we calculate that the mass of the atmosphere is 4.2×10^{23} gm, almost 100 times the mass of the terrestrial atmosphere.

On Mars a surface pressure of .006 atmospheres has been deduced from

---

*It is common practice to refer to this quantity as one *atmosphere* of pressure.
†We have made no allowance in this computation for the volume of land above sea level.

data obtained by Mariner spacecraft as well as from measurements using ground-based telescopes. With this information we calculate that the mass of the atmosphere is 2.4×10^{19} gm. There is no indication that Mercury has any atmosphere at all, and on the outer planets, as we have already noted, no clear demarcations exist between atmospheres and interiors.

Physical Data for the Planets

Now that we have discussed the compositions and masses of the atmospheres of the planets, we can turn our attention to the external factors that influence atmospheric properties and behavior (see Table 1–5). We have already noted that the gravitational acceleration is important because an atmosphere is a compressible fluid bound to its planet by the force of gravity.

The distance of the planet from the Sun determines the flux of solar radiation incident on the planet, and thus, as we shall show in Chapter 3, the average temperature. Imagine a series of spherical shells centered on the Sun, with radii equal to the distances of the planets from the Sun (Fig. 1–5). The amount of solar radiation passing through each shell is the same, but the surface area of the shells increases as the radius of the shells increases. In fact, it increases as the square of the radius, because the area of each spherical surface is $4\pi a^2$, where a is the radius of the sphere. This means that the flux of solar radiation, which is the amount of radiation passing through unit area of surface in unit time, must be inversely proportional to the square of the distance from the Sun. Since we know the flux of solar radiation at the Earth, we can calculate the flux for any other

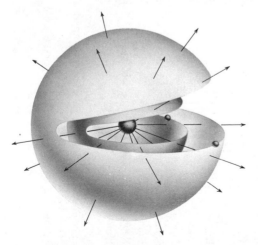

FIGURE 1-5 *Radiation crossing successive centered shells (cut away for illustration). The same amount of radiation must cross each shell. The flux, therefore, must decrease as the area of the shells becomes larger.*

The sun and the planets

Table 1-5

Physical Data for the Planets

Planet*	Mass (10²⁶ gm)	Mean Radius (km)	Mean Density (gm cm⁻³)	Gravitational Acceleration (cm sec⁻²)	Average Distance from Sun (10⁶ km)	Length of Year (Days)	Inclination of Equator to Orbit (Degrees)	Orbital Eccentricity	Period of Rotation (Days)
Mercury	3.35	2439	5.51	376	58	88	(0)†	.206	58.7
Venus	48.7	6049	5.26	888	108	225	<3	.007	−243‡
Earth	59.8	6371	5.52	981	150	365	23.5	.017	1.00
Mars	6.43	3390	3.94	373	228	687	25.2	.093	1.03
Jupiter	19,100	69,500	1.35	2620	778	4330	3.1	.048	0.41
Saturn	5690	58,100	0.69	1120	1430	10,800	26.8	.056	0.43
Uranus	877	24,500	1.44	975	2870	30,700	98.0	.047	−0.89‡
Neptune	1030	24,600	1.65	1134	4500	60,200	28.8	.009	0.53
Pluto	11	—	—	—	5900	90,700	—	.247	(6.39)†

*The first four planets are similar in size, mass, density, and probably chemical composition. They are the *inner planets*. The remaining five are very different from Earth, but apart from Pluto, are similar to one another. They are the *outer planets*.
†Data in parentheses are uncertain.
‡Venus and Uranus rotate in the opposite sense to the other planets.

(Data from W. M. Kaula, 1971, and S. H. Dole, 1970.)

planet using the distances of the Earth and the planet from the Sun. On Jupiter, for example, the solar flux is $(150/778)^2 = .037$ times the solar flux on Earth.

The length of the year, which we list in Table 1–5, is one of the fundamental time scales impressed on the atmosphere; as the planet travels around the Sun the seasons repeat themselves every year, and atmospheric behavior varies with the seasons.

The two quantities that determine the size of seasonal variations are inclination and the eccentricity of the orbit. The most important is the *inclination*, that is, the angle between the equator of the planet and the orbit (see Fig. 1–6). Imagine a planet for which this angle is 90°, as is nearly the case for Uranus. The axis of rotation of a planet always points in the same direction. Thus, at one stage of the planet's revolution around the Sun the north pole points toward the Sun and, to an observer on that planet, the Sun is directly overhead at the north pole; the entire southern hemisphere is then in darkness. When the planet reaches the opposite side of its orbit, the northern hemisphere is in darkness, and the southern hemisphere enjoys continuous daylight. The equator of Jupiter, on the other hand, lies very nearly in the orbital plane. Seasonal changes are small because the Sun is never farther than about 3° from the equator. The Earth and most of the other planets provide examples of intermediate inclinations. Seasonal variations are substantial, but continual winter darkness and summer daylight occur only in the neighborhood of the poles (broken lines mark the limits of the polar regions in the center panel of Fig. 1–6).

The second source of seasonal change is the *eccentricity of the orbit* around the Sun, that is, the extent to which the orbit deviates from a circle. Mercury and Pluto have relatively large eccentricities, which means that they are substantially closer to the Sun at perihelion than at aphelion (see Fig. 1–7). They are, therefore, exposed to stronger solar radiation at perihelion. The effect is not very large for the other planets, but it is interesting in the way it combines with the larger seasonal variation caused by inclination. The Earth, for example, passes through aphelion during northern summer and through perihelion during winter. The solar radiation falling on the Earth is, therefore, more intense during northern winter than during summer, which reduces the seasonal change. However, the situation is reversed in the southern hemisphere, where the seasonal change due to orbital eccentricity augments the seasonal change due to equatorial inclination. If all other things were equal, southern winter would be colder than northern winter and southern summer would be hotter than northern summer. On Mars the same effect allows the northern polar cap to persist through the summer while the southern polar cap disappears.

The period of rotation, or time a planet takes to turn once on its own axis, is another fundamental time as regards planetary atmospheres. In Chapter 4 we shall show that rotation has an important influence upon the

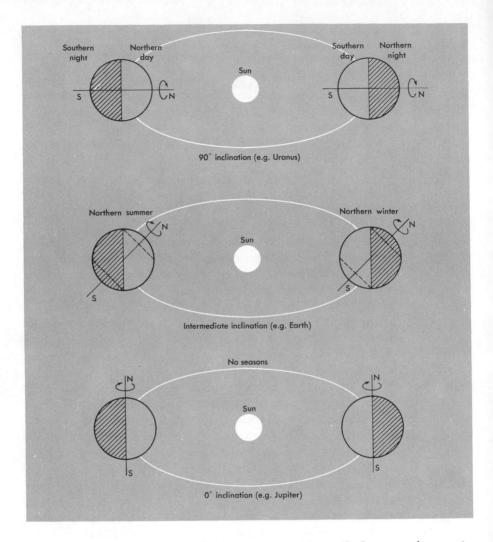

FIGURE 1-6 *How seasonal variations depend on the angle between the equator and the plane of the orbit. If a planet had exactly 90° inclination, it would be impossible to draw an analogy with terrestrial north and south poles. The labels in the top panel would then be arbitrary.*

flow of the atmosphere over the surface of the planet. The period of rotation is also related to the length of the day, but with a complication. To an observer on the planet, the yearly rotation around the sun is equivalent to the daily path of the sun through the sky. In effect, the observer gains or loses one sunset and sunrise per year depending on the relationship between the two directions of rotation.

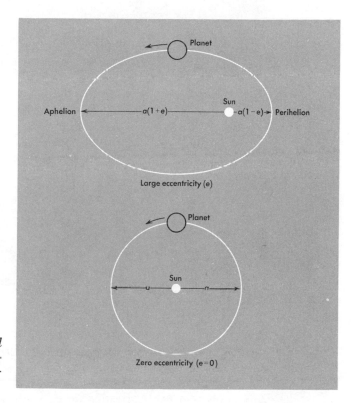

FIGURE 1-7 *How orbital eccentricity causes seasonal changes in the distance from the sun.*

Figure 1–8 illustrates the effect of orbital motion on the length of the day. Suppose that a planet with a marked side rotates on its own axis in the same direction and at the same angular velocity as it rotates in its orbit around the sun. The net result is that the marked side stays in permanent daylight and has a day that is infinitely long. We must allow for the fact that planets that rotate in the same direction around the sun and their own axes lose one rotation in each year. On the other hand, Venus, which rotates in the direction opposite to its orbital motion around the sun, gains one rotation. When the day is much shorter than the year, the difference between the length of the day and the period of rotation is negligible for our purposes. Only for Venus and Mercury is the difference significant. Venus has a day equal to 117 Earth days and Mercury has a day equal to 176 Earth days.

The Sun

The Sun is the major energy source for all the inner planets and is an important energy source for all the planets in the solar system. In order to

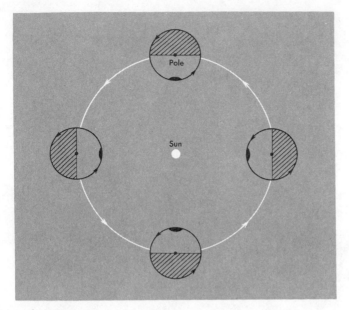

FIGURE 1-8 *The effect of orbital motion on the length of the day. The planet shown rotates synchronously in its orbit and around its axis. The day and night, therefore, last forever. The moon behaves in this manner in its motion around the Earth and always presents us with the same face.*

understand the planets, therefore, we must understand the solar environment, and in order to do that, we must consider some of the properties of the Sun itself.

The Sun emits two kinds of radiation that affect planetary atmospheres. There is a continual stream of electrically charged particles, the *solar wind*, which interacts with the very outer layers of planetary atmospheres. However, the overwhelming preponderance of the energy that travels from the Sun to the planets is carried by electromagnetic radiation. It is the effect of this energy on atmospheres that we wish to examine. Let us, therefore, consider the Sun's electromagnetic radiation, which consists of a mixture of radiations having different wavelengths (see Fig. 1–9).

The way in which a particular radiation interacts with an atmosphere depends on its wavelength and on the flux of radiant energy. The relationship between flux and wavelength for solar radiation is shown in the spectrum in Fig. 1–10, which extends from less than 10^{-7} cm to more than 10^3 cm. The very short waves are X rays, and the very long waves are radio waves, but different names do not indicate any fundamental differences in the nature of the waves. The names only provide a convenient designation for the various wavelength regions of the electromagnetic spectrum.

Figure 1–10 shows that solar radiation is most intense in the visible region of the spectrum, with the flux decreasing towards longer and shorter wavelengths. This behavior, with a maximum in the flux at intermediate wavelengths, is characteristic of the radiation emitted by incandescent ma-

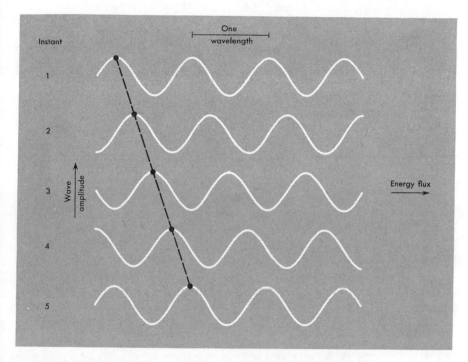

FIGURE 1-9 *Electromagnetic radiation as a wave motion. The same wave is shown at five successive instants of time. The dashed line connecting points of similar amplitude on the wave shows the wave moving to the right as time passes. The magnitude of the wavelength can vary from 10^{-6} cm and shorter for X rays to 10^5 cm and longer for radio waves. Infrared and visible radiation is often measured in microns (symbol μ, 10^{-4} cm) and visible, ultraviolet, and X rays in Angstrom units (symbol Å, 10^{-8} cm). Thus, red light has a wavelength of 0.7 μ or 7000 Å.*

terial. Experiment and theory have shown that both the wavelength at the maximum and the flux of the emitted radiation depend on the temperature of the radiating material. As the temperature is increased, the flux increases, and the wavelength at which it is greatest decreases (see Fig. 1–11). These phenomena may be observed in a fire or a furnace. The cold parts of the fire seem not to be radiating at all, but they are, however, radiating weakly in the infrared, where the eye is not sensitive. Warmer parts of the fire glow a dull red. The hottest parts of the fire glow with a bright white light. The radiation they emit is both stronger and more concentrated at short wavelengths than the radiation from the colder parts of the fire.

Measured spectra do not exactly correspond to the idealized curves shown in Fig. 1–11. Nevertheless, these idealized spectra, called *Planck spectra*, provide a very close description of the radiation from incandescent

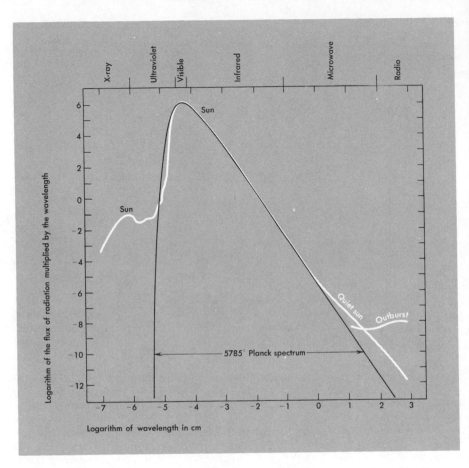

FIGURE 1–10 *Comparison between the spectrum of the Sun and a Planck spectrum corresponding to a temperature of 5785° K. (From C. W. Allen, 1958, Quart. J. Royal Meteorological Soc.)*

material under many conditions. In Fig. 1–10, for example, we see how closely the solar spectrum corresponds to the Planck spectrum for material at a temperature of 5785°K. There is a narrow region in the near ultraviolet where the solar spectrum falls below the Planck spectrum. In the microwave and radio regions, as well as in the far ultraviolet and X-ray regions, the solar spectrum rises above the Planck spectrum; but overall there is a close fit between the two spectra in the visible and infrared regions, where sunlight is strongest.

The deviations of the solar spectrum from the Planck spectrum may be understood as consequences of the variation of temperature with height in the outer layers of the Sun. The strongest solar radiation comes from a 300

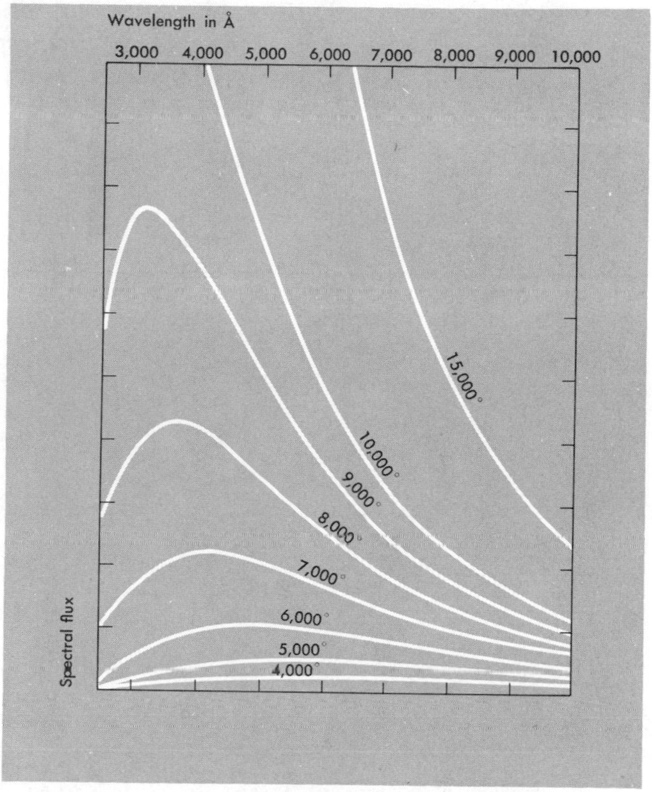

FIGURE 1-11 *The variation of the Planck spectrum with the temperature of the radiating material. As the temperature increases, the flux increases at all wavelengths, and the wavelength of the maximum decreases. (From Cecilia Payne-Gaposchkin and Katherine Haramundanis, Introduction to Astronomy, 2nd ed., 1970. By permssion of Prentice-Hall, Inc.)*

km thick layer in the solar atmosphere called the *photosphere*. Temperatures in the photosphere decrease from about 10,000°K at the bottom to 5,000°K at the top. Overlying the photosphere is a layer approximately 15,000 km thick, called the *chromosphere*. Temperatures increase from 5,000°K at the bottom of the chromosphere to 500,000°K at the top. Higher still is a region of the solar atmosphere called the *corona*, which tapers off gradually into interplanetary space. Temperatures in the corona are high and very variable. Their average value may be several million degrees.

Solar radiation at very short and at very long wavelengths comes from the upper chromosphere and the corona. The solar spectrum at these wavelengths is relatively enhanced because temperatures in the upper levels of the solar atmosphere are much larger than temperatures in the photosphere.

The properties of the photosphere do not vary importantly with time. Since the most intense solar radiation, in the visible and infrared regions of the spectrum, comes from the photosphere, there is little variability in the total rate at which the Sun radiates energy to a planet. There are, however, quite marked variations with time in the solar spectrum at very long and

The sun and the planets

very short wavelengths. These variations arise because of large changes of density and temperature in the upper chromosphere and corona. The atmospheres of the planets are affected in their upper levels by changes in the flux of ultraviolet and X radiation from the Sun.

2

Solar radiation

and chemical change

In Chapter 1 we set the stage for our discussion of planetary atmospheres by describing the overall properties of the atmospheres, particularly their masses and compositions, and by describing the solar radiation to which the atmospheres are exposed. In the remaining chapters we shall consider what happens as a result of the interaction of the atmospheres and the solar radiation. For a start, let us consider the chemical changes that take place when an atmosphere absorbs solar radiation at ultraviolet wavelengths.

Nitrogen, the most abundant gas in the Earth's atmosphere, suffers very little chemical change; most of the chemical reactions in the Earth's atmosphere involve oxygen. In the atmosphere, most oxygen exists in the form of molecules made up of two atoms bound together by chemical forces (the chemical symbol is O_2). Important changes, however, take place at high altitudes where the oxygen molecules absorb ultraviolet radiation from the Sun. What happens when a molecule absorbs radiation depends on the nature and quantity of energy absorbed.

Electromagnetic radiation may be thought of as a stream

of particles, called *photons*. The energy of each photon is inversely proportional to the wavelength of the radiation. Long wavelength photons, therefore, have little energy. When they are absorbed, they may cause the absorbing molecule to rotate or to vibrate, but they cannot cause chemical changes. To do that, photons must have enough energy to disrupt the molecule by, for example, breaking the chemical bond that holds the atoms together (see the third line in Fig. 2–1). This process, in which the absorption of a photon leads to disruption of the absorbing molecule, is called *photodissociation*. The atoms produced by photodissociation are chemically very active, and they start chains of reactions leading to the production of new species. Two new species produced in the Earth's atmosphere by the photodissociation of molecular oxygen (O_2) are atomic oxygen and ozone. As the name suggests, atomic oxygen consists of independent oxygen atoms that are not chemically bound (the chemical symbol is O). Ozone, on the other hand, consists of three oxygen atoms bound together in a single mole-

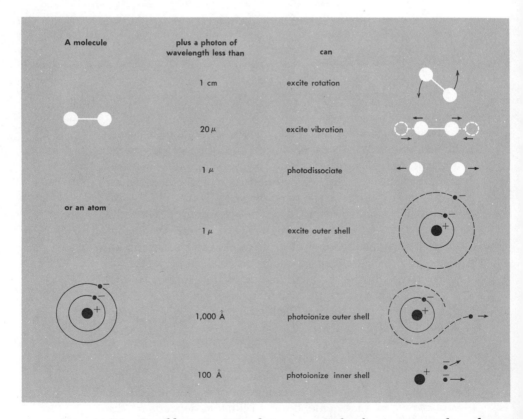

FIGURE 2-1 *Possible interactions between a molecule or atom and a photon. The longer wavelength events, which involve less energy, are at the top.*

Solar radiation and chemical change

cule (the chemical symbol is O_3). The abundance of these different forms of oxygen varies with altitude in the manner shown by the solid lines in Fig. 2–2. We must now seek a theoretical understanding of the vertical distributions of atomic and molecular oxygen and of ozone.

Chapman Profile

The chain of reactions that produces and destroys atomic oxygen and ozone starts with photodissociation of molecular oxygen. Before we can make further progress, therefore, we need to know how rapidly photodissociation occurs at different levels of the atmosphere.

Three factors determine the rate of photodissociation. The first is the number of photons available to cause photodissociation; the second is the number of oxygen molecules exposed to dissociating radiation; the third is the efficiency with which the photons in question cause dissociation. This efficiency depends on the wavelength of the photons. As an example, remember that long wavelength photons do not have enough energy to break the molecular bond, that is, to dissociate the molecule. On the other hand, we have learned from laboratory measurements of the absorption of radia-

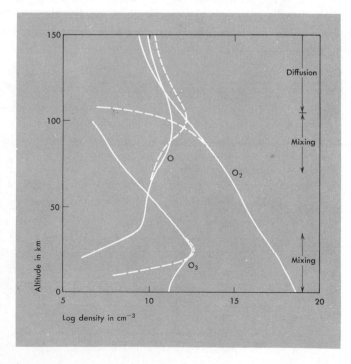

FIGURE 2-2 *Altitude profiles of the densities of atomic oxygen (O), molecular oxygen (O_2), and ozone (O_3) in the Earth's atmosphere. The broken lines show theoretical profiles calculated on the assumption of photochemical equilibrium. The regions in which diffusion and mixing cause departures from photochemical equilibrium are indicated on the right.*

Solar radiation and chemical change

tion by gases that photodissociation is not probable at very short wavelengths either. Molecular oxygen, for example, is most strongly dissociated by radiation at wavelengths near 1450 Å. At longer and at shorter wavelengths the efficiency of photodissociation decreases.

The efficiency with which photons cause photodissociation depends on the wavelength of the photons, but usually not on the height in the atmosphere. The other two factors, however, the density of oxygen molecules and the amount of dissociating radiation, do change as the height changes. The variation of the oxygen density is given by the barometric law discussed in Chapter 1. The number of molecules in each cubic centimeter of air decreases by a factor of e ($= 2.7182$) for each increase in altitude of one scale height. This variation of density with altitude is illustrated in Fig. 2–3(a). We see that, if other things were equal, more photodissociation would occur at low altitudes than at high altitudes because there are more molecules at low altitudes to be dissociated.

Other things are not equal, however. There is more dissociating radiation at high altitudes than at low altitudes. The quantity of interest here is the flux of radiation, or the number of photons flowing through an area of one square centimeter every second. At altitudes well above the atmosphere the flux of dissociating radiation depends on the brightness of the Sun and on the distance from the Sun, as we described in Chapter 1. But a new factor is introduced within the atmosphere: photons are absorbed by the atmospheric gases, which means that the flux of photons becomes smaller as the radiation penetrates deeper into the atmosphere.

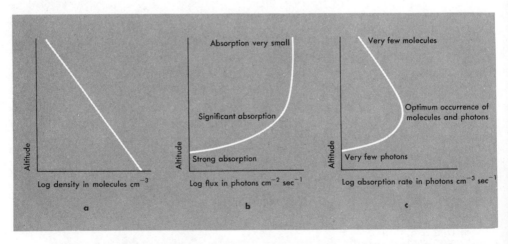

FIGURE 2-3 *The Chapman profile. The rate of absorption of solar photons is proportional to the product of the density of absorbing molecules and the flux of photons (the logarithmic scales add: $c = a + b - constant$)*

The rate at which the photon flux diminishes as the altitude decreases depends on the efficiency with which radiation is absorbed by the atmospheric gases. This efficiency, as we have already noted, depends on the wavelength of the radiation. For example, visible radiation, with wavelengths around 5000 Å, suffers little absorption in the atmosphere, and the flux at the ground is not much smaller than the flux in space. On the other hand, the atmosphere strongly absorbs ultraviolet radiation with wavelengths shorter than 3000 Å. Photons at these wavelengths do not reach the ground at all. For this reason we were able to learn little about the Sun's ultraviolet spectrum before rockets became available for scientific work in the late 1940's.

In Fig. 2–3(b) we show how flux decreases as altitude decreases for radiation that is absorbed in the atmosphere. At high altitudes there is little atmosphere and therefore little absorption; the flux is almost constant. As the atmospheric density increases, however, and absorption becomes increasingly important, the flux drops off more and more rapidly.

Now we can see how the rate of absorption of photons varies with altitude. This rate, which is proportional to the product of the density of absorbing molecules and the flux of photons that can be absorbed, is shown in Fig. 2–3(c). The rate is small at high altitudes because there are few absorbing molecules; the rate is small at low altitudes because there are few photons to be absorbed; the rate therefore has a maximum at intermediate altitudes, as the figure shows.

This description of the absorption of solar radiation in a planetary atmosphere was worked out in quantitative form by Chapman in 1930; the profile of absorption rate as a function of altitude is called the *Chapman profile*. As our discussion has indicated, the altitude of the maximum of the Chapman profile depends on the efficiency with which atmospheric gases absorb radiation, and this in turn depends on the wavelength of the radiation. Using Chapman's theory and laboratory measurements of absorption by atmospheric gases at different wavelengths, we can calculate, at each wavelength, the height at which the rate of absorption of solar radiation is a maximum. Results of such a calculation are shown in Fig. 2–4 for wavelengths shorter than 3200 Å. At longer wavelengths, in the visible and near infrared spectrum, most of the solar radiation is absorbed at the ground or very close to the ground (see Fig. 3–4 for a different view of the same phenomenon).

Photodissociation of Molecular Oxygen

Figure 2–4 shows that photodissociation of molecular oxygen occurs mainly in the wavelength region between 1000 Å and 2000 Å and that the

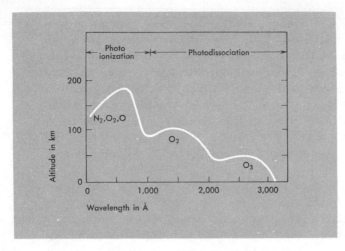

FIGURE 2-4 *Altitude in the Earth's atmosphere, as a function of wavelength, at which the rate of absorption of solar radiation is a maximum. The absorbing molecules are indicated below the curve and the absorption processes are shown above. Photodissociation causes disruption of the absorbing molecule. Photoionization causes the ejection of an electron from either a molecule or an atom, leaving behind a positively charged ion (see Fig. 2-1).*

rate of photodissociation is a maximum at a height near 100 km. We can write the process of photodissociation as a chemical reaction between an oxygen molecule and a photon of ultraviolet radiation, producing two oxygen atoms

$$O_2 + \text{photon} \longrightarrow O + O \tag{2-1}$$

Our discussion shows that oxygen atoms are produced most rapidly by this process near 100 km. We already have some understanding, therefore, of why atomic oxygen is abundant at high altitudes, but not at low altitudes, as shown in Fig. 2-2.

There must be more to the story of atomic oxygen. If the only process were photodissociation of molecular oxygen, we would in time find that all of the oxygen in the atmosphere would be converted from the molecular form to the atomic. What is missing is a process that recombines atomic oxygen to form molecular oxygen.

There are several ways for this recombination to take place. Atomic oxygen, which is highly reactive, can combine with many molecules and atoms present in the atmosphere. The resulting molecules can in turn react to produce the molecular oxygen needed to replace that which has been photodissociated. One example, which we shall discuss later, is the reaction between atomic and molecular oxygen to form ozone. This and other reactions have to be considered to account for the balance of atomic and molecular species at all levels in the atmosphere. As a first step in our discussion, however, it is useful to consider only the simplest and most direct form of recombination, namely that between two atoms of oxygen to form one oxygen molecule. Our conclusions, based on this recombination reaction alone, will have some relevance at altitudes above 60 km.

We can write this recombination reaction in the form

$$O + O + M \longrightarrow O_2 + M \qquad (2\text{--}2)$$

where M represents any other atmospheric molecule. There is an important reason why this third molecule, M, has to take part in the reaction. Chemical energy is released when two oxygen atoms combine to form a molecule. If the oxygen molecule is unable to get rid of this excess energy, the molecule will come apart again, recreating the original two oxygen atoms (see Fig. 2–5). In reaction (2–2), the oxygen molecule gives up its excess energy in a collision with the molecule M before it falls apart.

Let us assume that the two reactions, (2–1) and (2–2), determine the balance between atomic and molecular oxygen in the Earth's atmosphere. The first reaction destroys molecular oxygen and produces atomic oxygen, while the second reaction removes atomic oxygen and restores the molecular oxygen. The two reactions must occur equally fast since, on the average, the ratio of atomic to molecular oxygen in the atmosphere is neither increasing nor decreasing with the passage of time.

The rate of the recombination reaction (2–2) depends on the densities of the constituents that take part in the reaction. The reaction can only take place when two oxygen atoms happen to collide with one another and

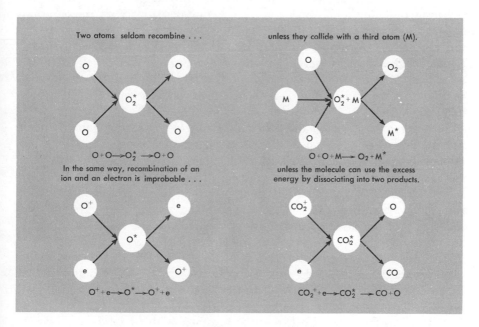

FIGURE 2-5 *Recombination does not occur in a gas unless there is a second product of the reaction to carry away the chemical energy released. The stars denote that the atoms or molecules have excess energy.*

Solar radiation and chemical change

with a third molecule, M, at the same time. Obviously collisions are much more frequent at high densities, when there are many atoms or molecules in each cubic centimeter, than at low densities, when the atoms and molecules are far part. The *law of mass action* expresses this consideration in quantitative form: the rate of a chemical reaction is proportional to the product of the densities of the species taking part in the reaction. For reaction (2–2) this means that the rate of recombination is proportional to (density of atomic oxygen) × (density of atomic oxygen) × (total density of atmospheric molecules).

Now that we know how the rate of recombination depends on the density of atomic oxygen we are in a position to calculate the way in which the density of atomic oxygen varies with altitude. As a first approximation, we can assume that a balance exists, at each altitude, between the rate of production of atomic oxygen and the rate of removal of atomic oxygen. In other words, we can equate the rate of reaction (2–2) to the rate of photodissociation, which is given by a Chapman profile resembling that shown in Fig. 2–3. The assumption of balance between production and loss at each altitude is known as the assumption of *photochemical equilibrium.* It leads to a theoretical profile of atomic oxygen density, shown by the broken line in Fig. 2–2, which increases as the altitude increases up to 100 km. This increase occurs partly because of the increasing flux of dissociating radiation and partly because recombination occurs more rapidly at lower altitudes where total atmospheric density is large (remember that the rate of recombination is proportional to total atmospheric density). Photochemical equilibrium calculations agree reasonably well with observed densities of atomic oxygen in the Earth's atmosphere for altitudes up to about 70 km, but as Fig. 2–2 shows, there are discrepancies at greater heights.

In order to understand these discrepancies we must look into the competitive processes operating in the atmosphere, each process trying to produce a different vertical distribution of chemical species. Up to this point we have discussed chemical and photochemical processes which, by themselves, would give the distribution appropriate to photochemical equilibrium. However, there are two other processes, *mixing* and *diffusion,* that strive to produce different distributions. Diffusion alone would give rise to an atmosphere in which each species is distributed according to a barometric law with its own characteristic scale height. We shall return to this phenomenon later in this chapter, but here we may note that below 100 km ozone and atomic oxygen clearly do not have the distribution with height associated with the barometric law (see Fig. 2–2).

Mixing has the same effect in air as in other fluids; it tends to cause the chemical species to have the same relative concentrations in each part of the fluid. If mixing were to dominate the profiles in Fig. 2–2, the curves for atomic and molecular oxygen and ozone would be parallel because of the logarithmic density scale. This is obviously not the case over most of the altitude range under consideration.

Now suppose we know that two out of three of the processes are to some extent involved and we wish to decide which one is the most important. The best indication is obtained by comparing the rates at which the processes take place. Suppose the atmosphere is initially far from mixed and that mixing processes are then allowed to act. If there is neither diffusion nor photochemistry, the distribution will in time tend to that appropriate to a strongly mixed atmosphere. We can estimate the approximate time it will take for this distribution to be reached. This time is the *time constant for mixing.* If the procedure were repeated with photochemistry or diffusion in place of mixing, we would find two additional time constants —one for photochemistry and one for diffusion.

Now, if two physical processes are competing and one acts more rapidly than the other, usually the faster one will be the more effective. The faster process is the one with the shorter time constant.

Let us consider the oxygen chemistry described by Eqs. (2–1) and (2–2). The rate of reaction of atomic oxygen in Eq. (2–2) depends, according to the law of mass action, on the total concentration of all kinds of molecules in the atmosphere, in the sense that the higher the concentration (that is, the lower the altitude), the greater the rate and the shorter the time constant.

Mixing, on the other hand, is a process that does not vary in such a simple way with height. Mixing is caused by winds and meteorological processes. These vary greatly from one place to another, but there is, on the average, no systematic change with height that compares with the rapid change in density.

We, therefore, anticipate that there is a height below which the photochemical time constant is shorter than the time constant for mixing and above which the reverse is true. Below this height we expect to find approximate photochemical equilibrium, as seen in Fig. 2–2; above there should be a tendency towards a mixed distribution. If we calculate the chemical time constant for reaction (2–2) at 90 km, we find it is about 100 days. Experience suggests that the time constant for atmospheric mixing is much less than 100 days, so we conclude that mixing is important at 90 km.

Mixing carries oxygen atoms downwards to lower levels where they can recombine rapidly. As a result, the density of atomic oxygen falls below the photochemical equilibrium density at altitudes above 90 km.

This behavior is illustrated in Fig. 2–2, where we see that the result is a density profile with a maximum at a height of about 95 km. The fraction of the oxygen that is dissociated is quite substantial. At 95 km the density of atomic oxygen is about one-twentieth of the density of molecular oxygen, and by 120 km the two densities are approximately equal.

The Influence of Diffusion

The ratio of atomic to molecular oxygen continues to increase as the altitude increases above 120 km (see Fig. 2–2), but this is due to diffusion. It is a result of the action of gravity, which exerts a stronger force on relatively heavy molecules, such as molecular oxygen, than on lighter constituents, such as atomic oxygen.

In order to understand the consequences of the differences in gravitational force, let us reexamine our discussion of the barometric law. We showed in Chapter 1 that atmospheric pressure and density decrease by a factor of e ($= 2.7182$) for each increase in altitude of one scale height. The scale height is given by (Eq. 1–8)

$$H = R \times \frac{\text{Temperature}}{\text{Gravitational acceleration}} \qquad (2\text{–}3)$$

where R, the gas constant, is inversely proportional to the mass of the gas molecules. Consequently, as we noted in Chapter 1, the scale height varies inversely with molecular mass, and densities of heavy gases, such as carbon dioxide, tend to fall off more rapidly with height than densities of light gases, such as hydrogen. On a given planet, however, this behavior is apparent only when diffusion controls gaseous concentrations.

This tendency for the atmospheric constituents to separate under the action of gravity exists at all altitudes, but its effects are not noticeable low in the atmosphere because the time constant for diffusion is longer than that for mixing. Mixing is therefore the more important process. The time constant for diffusion depends on the rate at which the heavier constituents can move downwards and the lighter constituents can move upwards. This flow of the different gases through one another is slowed down by collisions. Therefore, at low altitudes, where atmospheric density is high and collisions are frequent, the flow is much slower than at high altitudes where collisions are infrequent (see Fig. 2–6). In fact, above 120 km it takes only a day or so for diffusive separation to occur, but it would take about 30,000 years for the whole atmosphere to separate in this manner all the way down to the ground.

The atmosphere is turned over by winds and turbulence so many times in 30,000 years that diffusive separation is completely negligible compared to mixing near the ground. This means that, in the lower atmosphere, the chemical species are thoroughly mixed, and the densities of most atmospheric constituents decrease with altitude at the same rate (with a scale height that corresponds to the average molecular mass of the mixture of constituents). Diffusion, however, becomes more and more rapid as the

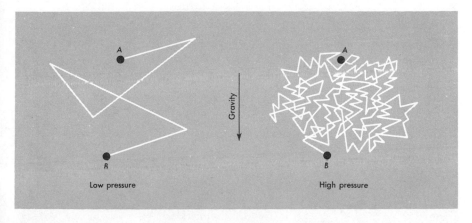

FIGURE 2-6 *When molecules collide, they change direction erratically. The path from A to B is much longer when collisions are frequent (right) than when collisions are infrequent (left).*

altitude increases and atmospheric density decreases. Inevitably, at sufficiently high altitudes, diffusion must be more important than mixing. On Earth the altitude at which diffusion becomes more important than mixing is about 105 km. Above 105 km the different constituents are able to separate. As a result, the density of each atmospheric constituent decreases at a different rate, with a scale height that corresponds to its own molecular mass. The heavier constituents, with smaller scale heights, fall off more rapidly than the lighter constituents, and the average molecular mass of the atmospheric gas decreases steadily as the altitude increases.

At 120 km, 80% of the atmosphere is molecular nitrogen (28 atomic mass units), and the average molecular mass is slightly higher than 28 amu. By 300 km, molecular nitrogen and oxygen have become much less abundant than atomic oxygen (16 amu), and the average molecular mass is close to 16 amu. At higher altitudes still, helium, with a mass of only 4 amu, becomes increasingly important, and at the top of the atmosphere, the most abundant constituent is the lightest gas of all, atomic hydrogen (1 amu).

Figure 2–7 presents representative density profiles for the different constituents in the upper atmosphere of the Earth, and it shows how the densities of the heavier constituents decrease more rapidly with altitude than the densities of the lighter constituents. The figure also shows how the density profiles depend on the temperature. The scale heights of the various constituents are proportional to temperature, as shown by Eq. (2–3). When temperatures are high, therefore, the densities of the different atmospheric species decrease slowly with altitude, which means that densities at high altitudes are relatively large. When atmospheric temperatures are low, on the other hand, the densities at high altitudes are also low. Temperatures

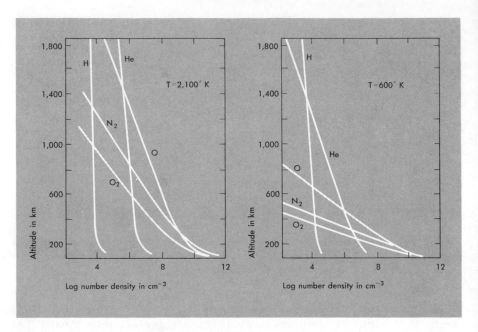

FIGURE 2-7 *Density profiles for the different constituents in the upper atmosphere of the Earth for two different values of the atmospheric temperature. (From J. C. G. Walker, 1964, Space/Aeronautics.)*

in the upper atmosphere vary through a wide range, as we shall describe in Chapter 3.

The same phenomenon of diffusive separation occurs in the upper atmospheres of Venus and Mars, and we might expect the light species, atomic oxygen, to be relatively abundant at high altitudes in these atmospheres as it is in the atmosphere of the Earth. Measurements show that this is not so, although the reasons are not known. How the relative abundance of atomic oxygen in the upper atmospheres of the planets affects the structure of their *ionospheres* is the subject of the next section.

Photoionization and Planetary Ionospheres

The terrestrial ionosphere is a region of the atmosphere, extending upwards from a height of about 50 km, in which electrically charged ions and electrons are present in sufficient quantities to affect the propagation of radio waves. Most of the ions and electrons are produced when electromagnetic radiation from the Sun is absorbed by uncharged atmospheric molecules. As shown in Fig. 2–1, only photons with wavelengths less than about 1000 Å have enough energy to knock an electron out of an atmospheric molecule (to *photoionize* the molecule), so it is the short wavelength, extreme ultraviolet radiation, that is responsible for the formation of the ionosphere. Photoionization occurs high in the atmosphere because atmospheric gases absorb strongly in the extreme ultraviolet region of the spectrum (see Fig. 2–4).

The situation is similar on all the planets. On Venus, for example, the rate of photoionization is greatest at an altitude of about 140 km (see Fig. 2–8). Since carbon dioxide is the most abundant gas at this altitude, most of the ions that are produced are carbon dioxide ions, CO_2^+,

$$CO_2 + \text{photon} \longrightarrow CO_2^+ + \text{electron} \qquad (2\text{–}4)$$

Experience indicates that molecular ions such as CO_2^+ generally recombine rapidly with electrons in a reaction that neutralizes the ion and the electron and dissociates the molecule. For CO_2^+, the reaction produces carbon monoxide and atomic oxygen,

$$CO_2^+ + \text{electron} \longrightarrow CO + O \qquad (2\text{–}5)$$

The evidence available at this time indicates that the structure of the ionosphere on the sunlit side of Venus can be explained entirely in terms of photochemical equilibrium between the rate of production of CO_2^+ by reaction (2–4) and the rate of removal of CO_2^+ by reaction (2–5).

It is easy to calculate how the ion and electron density varies with alti-

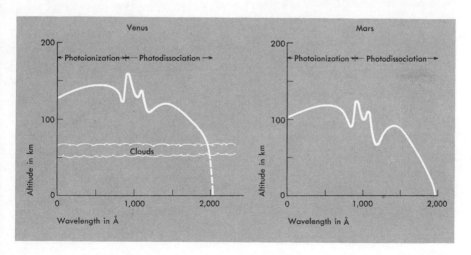

FIGURE 2-8 *The altitude in the atmospheres of Venus and Mars at which the rate of absorption of solar radiation is a maximum, as a function of the wavelength of the radiation.*

Solar radiation and chemical change

tude for such a simple system involving only two reactions. The rate of production of ions and electrons as a function of altitude is given by the Chapman profile described earlier in this chapter and shown again in Fig. 2–9 (left). For the rate of removal of ions and electrons we refer once more to the law of mass action. According to this law, the rate of the recombination reaction (2–5) is proportional to the product of the ion density and the electron density. The ion density and the electron density are almost exactly equal to each other, so we may take the rate of recombination to be proportional to the square of the density of electrons. We calculate the density by equating the rates of production and of recombination, and we find that the electron density is proportional to the square root of the photoionization rate. The constant of proportionality may be determined from laboratory measurements of the reaction between CO_2^+ and electrons. Results of such a calculation are shown in Fig. 2–9 (right), where they are compared with the electron densities actually measured on Venus by the Mariner V spacecraft.

We do not have to do any calculations to see that a similar theory cannot explain the properties of the Earth's ionosphere. Figure 2–4 shows that most of the photoionizing radiation, which has wavelengths shorter than 1000 Å, is absorbed at heights between 100 km and 200 km in the Earth's atmosphere. As a result, the rate of photoionization has its maximum at 150 km (see Fig. 2–10). Measurements of the Earth's ionosphere given in

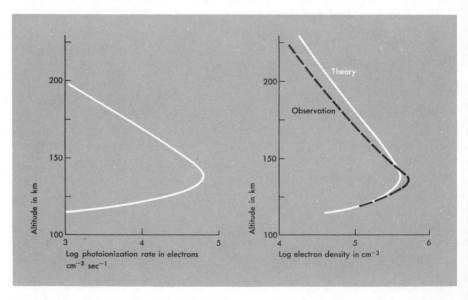

FIGURE 2-9   (Left) The photoionization rate as a function of altitude in the atmosphere of Venus. (Right) A comparison of calculated and measured electron densities. (After M. B. McElroy, 1968.)

Solar radiation and chemical change

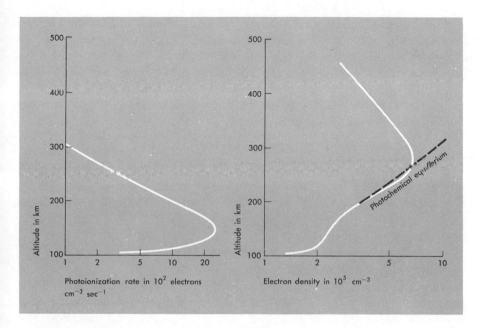

FIGURE 2-10 *(Left) The photoionization rate as a function of altitude in the upper atmosphere of the Earth. (Right) A typical profile of electron density. Also shown is the density profile calculated for photochemical equilibrium.*

Fig. 2–10 show, however, that the density of electrons has a maximum at a considerably greater height. This contradicts the simple photochemical theory, just developed for Venus, which predicts that the electron density is proportional to the square root of the photoionization rate, so that the maximum electron density should occur at the same height as the maximum photoionization rate.

The explanation lies in the nature of the ions formed by photoionization in the upper atmosphere of the Earth. Because atomic oxygen is the most abundant constituent of the atmosphere above 200 km (see Fig. 2–7) most of the ions produced above this height are atomic oxygen ions, O^+,

$$O + photon \longrightarrow O^+ + electron \qquad (2\text{–}6)$$

Consider now the recombination, remembering that it is almost impossible to combine two particles such as O^+ and an electron into one particle, such as O, in a reaction in which no other particles are involved. As we noted in our discussion of the recombination of atomic oxygen, a chemical reaction occurring in a gas must have at least two products to carry away the chemical energy released by the reaction (see Fig. 2–5). When atomic oxygen recombines at heights of around 100 km in the Earth's atmosphere, the excess energy is carried away by a third atmospheric molecule. At ionospheric heights, however, atmospheric density is much lower, and there are too few molecules present for reactions involving three constituents to be an important means for recombining ions and electrons. There is, therefore, no process of recombination between an atomic ion and an electron that is fast enough to be important in the ionosphere. Instead, the O^+ ions react chemically with neutral molecules of oxygen and nitrogen to produce molecular ions. The reactions are

$$O^+ + O_2 \longrightarrow O_2^+ + O \tag{2-7}$$

and

$$O^+ + N_2 \longrightarrow NO^+ + N \tag{2-8}$$

The molecular ions, once formed, are able to recombine with electrons by dissociation,

$$O_2^+ + \text{electron} \longrightarrow O + O \tag{2-9}$$

and

$$NO^+ + \text{electron} \longrightarrow N + O \tag{2-10}$$

thus destroying the ionization.

Let us apply the law of mass action to reactions (2–6), (2–7), and (2–8). The rate of creation of O^+ ions by reaction (2–6) is proportional to the product of the densities of O and of photons. However, we are above the maximum of the Chapman profile, and in this region of the ionosphere, the flux of photons does not vary much with height (see Fig. 2–3). Hence, the important feature of the production rate is that it is proportional to the density of atomic oxygen.

$$\text{Rate of production of } O^+ \propto O \text{ density} \tag{2-11}$$

Equations (2–7) and (2–8) enable us to calculate the destruction rate of O^+. It is proportional to the product of the density of O^+ and the density of O_2 or N_2,

$$\text{Rate of destruction of } O^+ \propto (O^+ \text{ density}) \times (O_2 \text{ or } N_2 \text{ density}) \tag{2-12}$$

If the O^+ density is not to vary with time, destruction and production rates must be equal. Consequently, the right-hand sides of Eqs. (2–11) and (2–12) must be proportional. By this argument we find

$$O^+ \text{ density} \propto \frac{O \text{ density}}{O_2 \text{ or } N_2 \text{ density}} \tag{2-13}$$

Now, we are in a region of the atmosphere in which diffusion controls the atomic and molecular densities. The O density, therefore, has a scale height corresponding to its atomic mass of 16 amu, while O_2 and N_2 have scale heights corresponding to 32 and 28 amu respectively. Thus, the ratio on the right-hand side of Eq. (2–13) increases continually as height increases (see Fig. 2–11). Since the atmosphere is neither positively nor negatively charged, and since O^+ is the most abundant ion, it is a good approximation to equate the O^+ density to the electron density, as indicated

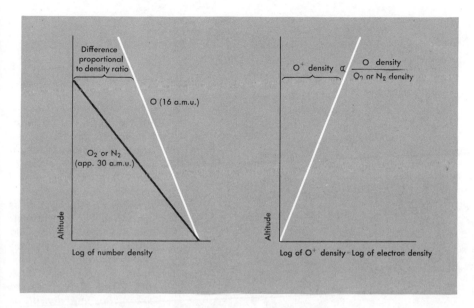

FIGURE 2-11 *Photochemical equilibrium for O^+ ions in the terrestrial atmosphere.*

on the right of Fig. 2–11. This mechanism accounts for the steady increase in ionization in the Earth's atmosphere below 300 km (see Fig. 2–10).

Our description of the photochemistry of the ionosphere does not yet explain why the density reaches a maximum at a height of about 300 km and decreases above. Indeed, our argument indicates that the density of ions and electrons should increase with altitude until all of the atmospheric gas is ionized. We must, however, consider the role of competitive, nonphotochemical processes.

When we discussed atomic oxygen, we found that photochemistry was first overcome by mixing, and later, at a greater height, mixing was overcome by diffusion. In the case of the ionosphere, mixing is not important. The time constant for the destruction of O^+ by reactions (2–7) and (2–8) is shorter than the time constant for mixing at all heights of importance so the competition is between photochemistry and diffusion (see Fig. 2–12). We have already shown that the time constant for diffusion decreases with height and that this process is bound to dominate at sufficiently large altitudes. Because the photochemical step proceeds very rapidly, however, it is only at low pressures or high altitudes that diffusion can dominate. There is no contradiction involved in identifying 300 km as the level for diffusion to control ion densities, although we identified 105 km for the same event when we discussed atomic oxygen; this simply reflects the fact that the reactions of O^+ ions proceed much more rapidly than the recombination of atomic oxygen.

Solar radiation and chemical change

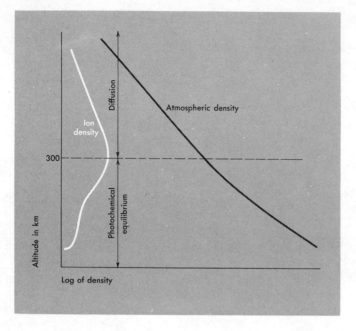

Altitude in km

Log of density

FIGURE 2-12 *At altitudes below 300 km the atmospheric density is relatively high. Diffusion is therefore slow, and chemical reactions are fast. The ion density in this region is controlled by photochemical reactions. Above 300 km the atmospheric density is relatively low. Diffusion is therefore fast, and chemical reactions are slow; the ion density is controlled by diffusion.*

Above the diffusion level oxygen ions are distributed according to a barometric law. The effective molecular mass is 8 amu because the charged medium is a mixture of equal parts of O^+ (16 amu) and electrons (approximately 0 amu). The important point is that density must decrease with height and, therefore, a maximum exists where diffusion first becomes important.

One of the phenomena we described in Chapter 1 is also relevant to the discussion of planetary ionospheres. It is the variability of conditions in the outer layers of the Sun's atmosphere and the corresponding variability of the radiation emitted. The solar radiation that causes photoionization lies in the extreme ultraviolet region of the spectrum (wavelengths less than 1000 Å) and originates in the upper chromosphere and corona. Changes in these regions of the solar atmosphere, therefore, cause detectable variation in the density of electrons in the ionosphere.

Ozone

Figure 2–8 shows that solar ultraviolet radiation with wavelengths between 2000 Å and 3000 Å penetrates to the surface of Mars. On Earth, on the other hand, as Fig. 2–4 shows, this radiation is absorbed by ozone at altitudes as high as 50 km. Radiation at these wavelengths is lethal for terrestrial organisms, and we could not survive in the open if the surface of the Earth were not protected by a screen of ozone.

Ozone consists of three oxygen atoms bound together in a single molecule. Its formation follows the photodissociation of molecular oxygen by ultraviolet radiation from the Sun (Eq. 2–1),

$$O_2 + \text{photon} \longrightarrow O + O$$

We have already described how the oxygen atoms produced by this reaction are removed at altitudes above about 70 km by the recombination reaction (2–2) between two oxygen atoms and a third molecule. At lower altitudes another reaction is important, a reaction between an oxygen atom and an oxygen molecule to produce an ozone molecule,

$$O + O_2 + M \longrightarrow O_3 + M \qquad (2\text{–}14)$$

The third molecule (M) is required here, as in reaction (2–2), to carry away chemical energy released by the reaction.

There are a number of reactions that remove the ozone produced by reaction (2–14). The first reaction, important at high altitudes, is the recombination of ozone and atomic oxygen to give two oxygen molecules,

$$O + O_3 \longrightarrow O_2 + O_2 \qquad (2\text{–}15)$$

The second reaction, which is important at all altitudes, is photodissociation of ozone,

$$O_3 + \text{photon} \longrightarrow O_2 + O \qquad (2\text{–}16)$$

This is the process responsible for the absorption of the solar radiation between 2000 Å and 3000 Å. Finally, ozone is removed at low altitudes by reactions with dust particles or by reactions with the ground.

Because of the number of reactions involved, ozone chemistry can be confusing. A rough analysis, however, gives us an idea of the behavior to expect. We can anticipate that there is little ozone at high altitudes because the reaction (2–14) that forms ozone is slow at altitudes where the densities of O_2 and of M are low. On the other hand, the formation reaction requires atomic oxygen (O) as well, and there is little atomic oxygen at low altitudes where dissociating solar radiation does not penetrate (see Fig. 2–2). We anticipate, therefore, that ozone densities are low in the lower atmosphere as well as at high altitudes. The maximum ozone density should occur somewhere between these two extremes. The detailed theory shows, and observation confirms, that on the average the maximum occurs at about 25 km.

In spite of its important role as an absorber of solar ultraviolet radiation, there is very little ozone in the atmosphere. If the whole amount were concentrated in a layer of pure ozone and compressed to one atmosphere pressure, the layer would be less than half a centimeter thick. Measurements show that the amount of ozone in the atmosphere varies substantially from day to day, and that there are also regular variations with season and latitude. These regular variations are shown in Fig. 2–13, where we see that ozone amount is a maximum at polar latitudes in the spring and a minimum at the equator in the winter. These results are surprising because a theory based on photochemical equilibrium between the production and removal of ozone predicts that ozone content is at a maximum at the latitude of the Sun, where the radiation flux and production rate are maximal.

Deviations from photochemical equilibrium for ozone occur for the same reasons as deviations from photochemical equilibrium for atomic oxygen. At altitudes below about 25 km, ozone molecules react so slowly that they can be carried by winds and turbulence far from the region of the atmosphere where they are formed (see Fig. 2–2). The atmospheric circulation is such that it transports ozone during the winter from low latitudes to the poles. This transport is responsible for the spring maximum at high lati-

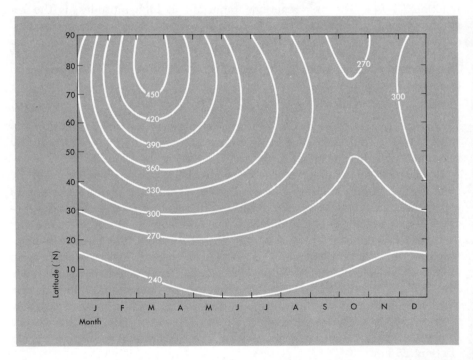

FIGURE 2-13 *The total amount of ozone in the atmosphere as a function of month and of latitude, averaged over a period of several years in the northern hemisphere. The lines connect points of equal ozone amount. The units give the thickness, in thousands of a centimeter, of a layer composed of all of the ozone in the atmosphere compressed to 1 atmosphere pressure. (From W. L. Godson, 1960, Quart. J. Royal Meteorological Soc.)*

Solar radiation and chemical change

tudes. Transport is less rapid during the summer, and the maximum gradually decays and moves to lower latitudes, as Fig. 2–13 shows.

The long photochemical lifetime of ozone molecules at altitudes below 30 km is unexpected. It is not that the molecules are not being destroyed —ozone molecules are dissociated by solar radiation in the visible and near infrared regions of the spectrum at altitudes all the way down to the ground; rather, it is that, at the lower altitudes, almost all of the oxygen atoms released by photodissociation of ozone (2–16) react immediately with an oxygen molecule (2–14) to form ozone. In this case, no net loss of ozone results from photodissociation, and the ozone behaves almost as an unreactive component of the air. This is why the ozone distribution is sensitive to circulation and why ozone measurements can be used to trace movement of the air.

It is not as easy to measure the altitude profile of the ozone density as it is to measure the total amount of ozone in the atmosphere. Enough measurements, however, have been made to show that the ozone density is close to photochemical equilibrium at altitudes above 40 km. The large variations caused by atmospheric circulation are restricted to the lower altitudes where, as we have just explained, the photochemical lifetime of an ozone molecule is long.

A typical altitude profile is shown in Fig. 2–2, where we see that, even at the maximum, ozone constitutes only a minute fraction of the atmosphere. Ozone is important, in spite of this, because it absorbs solar ultraviolet radiation so strongly. We have already referred to the role of ozone in shielding the surface of the Earth. Ozone also significantly affects the thermal budget of the atmosphere. The energy of the solar radiation absorbed by ozone causes the temperature of the atmosphere to rise to a local maximum at a height of about 50 km. This is a phenomenon that we shall consider in the next chapter.

3

Atmospheric temperatures

In Chapter 2 we considered some of the chemical changes that occur when solar radiation is absorbed by the atmosphere of a planet. Chemical changes are only a part of the story, however. The energy carried by solar radiation affects the temperature of an atmosphere as well as the chemical composition.

The heating effect of solar radiation can be seen quite readily in Fig. 3–1, which shows how temperature varies with altitude at middle latitudes in the Earth's atmosphere. There are three maxima in the temperature profile; the first is at the ground, where the temperature is about 290°K; the second is at the level called the *stratopause*, at a height of 50 km, where the temperature is about 280°K; and the third occurs at heights above 200 km, in the layer of the atmosphere called the *thermosphere*, where the temperature can rise to 1000°K or more. On the right of the figure there is an indication of the height in the atmosphere at which solar radiation of various wavelengths undergoes absorption. This is a simplified representation of Fig. 2–4, which shows the altitude of absorption plotted against wavelength.

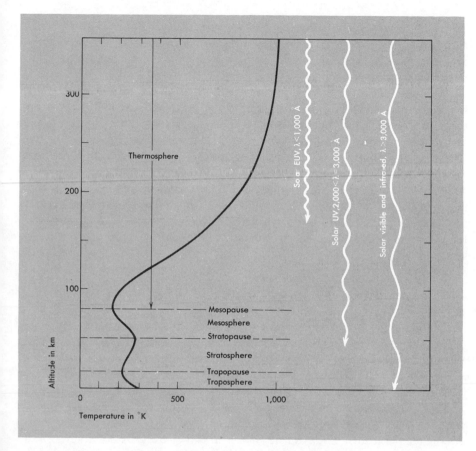

FIGURE 3-1 *The variation of temperature with altitude at middle latitudes in the Earth's atmosphere. Layers of the atmosphere are called troposphere, stratosphere, mesosphere, and thermosphere; these are separated by features called tropopause, stratopause, and mesopause. The high temperature regions of the atmosphere result from the absorption of solar radiation at different wavelengths, as shown on the right.*

The high temperature in the thermosphere is caused by solar heat, deposited at great heights as a result of the absorption of extreme ultraviolet radiation. The absorption process is photoionization, which we discussed in Chapter 2 as the source of the ionosphere. Similar processes of photoionization occur high in the atmospheres of all the planets, and we may anticipate that relatively hot thermospheres are a feature of all planetary atmospheres.

The second hot layer in the Earth's atmosphere, at a height of about 50 km, is the result of the absorption, by ozone, of solar ultraviolet radia-

Atmospheric temperatures

tion with wavelengths between 2000 Å and 3000 Å. This is the process, mentioned in Chapter 2, that dissociates ozone. Similar hot layers probably do not occur in the atmospheres of the other planets, because they lack significant amounts of ozone.

The third hot layer, which lies at the bottom of the atmosphere, results from absorption, largely by the ground, of solar radiation at visible and infrared wavelengths. We shall give most of our attention to this lower layer of the atmosphere, the *troposphere*, because this is where we live. Our goal is to understand the physical processes that determine the temperature of the ground and the temperature of the atmosphere near the ground. As we develop this understanding we shall find it helpful to compare conditions on Earth with conditions on the other planets, particularly the inner planets, Mars and Venus. A successful theory of atmospheric temperature should work on any planet.

Effective Temperatures of the Planets

It is generally believed that the planets are in a relatively steady state; that is to say, their temperatures are neither increasing nor decreasing. Averaged over a period of many years, therefore, the rate at which a planet radiates energy away to space must exactly balance the rate at which the planet receives energy from all sources.

The inner planets receive virtually all of their energy from the Sun, so we shall limit ourselves to consideration of how solar energy affects atmospheric temperatures. The rate at which a planet absorbs solar energy depends strongly on the distance of the planet from the Sun, because the flux of solar radiation—the amount of energy flowing across unit area in unit time (erg cm^{-2} sec^{-1})—varies inversely as the square of this distance (see Chapter 1). Table 3–1 shows the distance of each planet from the Sun and gives corresponding values of the solar flux. Each planet intercepts solar radiation at a rate (erg sec^{-1}) given by the product of the solar flux (erg cm^{-2} sec^{-1}) and the area of the disk of the planet (cm^2) as seen from the direction of the Sun (Fig. 3–2). This area is π times the square of the radius of the planet.

Not all of the radiation intercepted by a planet is absorbed. A fraction of the incident energy is reflected back to space by clouds and by the surface. This fraction is called the *albedo* of the planet. The albedo is different for every planet, depending on the nature of the atmosphere and surface. Values, determined from astronomical observations, are given in Table 3–1. Since the albedo is the fraction of the incident energy reflected by the planet, the fraction that is absorbed is (1 − albedo).

Table 3-1

Effective Temperatures

Planet	Distance from Sun (10⁶ km)	Flux of Solar Radiation (10⁶ erg cm⁻² sec⁻¹)	Albedo	T_e (°K)
Mercury	58	9.2	.058	442
Venus	108	2.6	.71	244
Earth	150	1.4	.33	253
Mars	228	.60	.17	216
Jupiter	778	.049	.73	87
Saturn	1430	.015	.76	63
Uranus	2870	.0037	.93	33
Neptune	4500	.0015	.84	32
Pluto	5900	.00089	.14	43

Combining these results, we find that a planet absorbs solar energy at a rate given by the following expression:

$$\text{Energy absorbed} = \pi \times (\text{Radius})^2 \times \text{Solar flux} \times (1 - \text{albedo}) \quad (3\text{–}1)$$

We must now consider the rate at which the planet radiates energy away to space because, as we have already noted, the rate of loss of energy must be equal to the rate of absorption of energy in the long run.

We described in Chapter 1 how hot material radiates energy more rap-

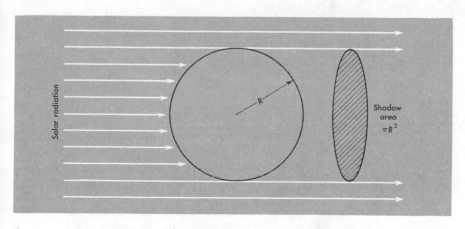

FIGURE 3-2 *From the direction of the Sun a planet looks like a disk with an area π times the square of the radius.*

Atmospheric temperatures

idly than cold material. According to the *Stefan-Boltzmann law*, the amount of heat energy radiated by a surface of unit area in unit time is proportional to the fourth power of the temperature (Fig. 3–3). We can write this law as

$$\text{Flux (erg cm}^{-2}\text{ sec}^{-1}) = \sigma \times (\text{Temperature})^4 \tag{3-2}$$

where σ is the constant of proportionality, called the *Stefan-Boltzmann constant* ($\sigma = 5.67 \times 10^{-5}$ erg cm^{-2} deg^{-4} sec^{-1}). Each unit area of planetary surface radiates energy at a rate given by Eq. (3–2), so to get the total rate at which the planet loses energy to space we must multiply the flux by the total surface area. The surface area of a sphere is 4π times the square of the radius, so we find

$$\text{Energy radiated} = 4\pi \times (\text{Radius})^2 \times \sigma \times (\text{Temperature})^4 \tag{3-3}$$

If we now equate the rate at which energy is radiated by the planet (Eq. 3–3) to the rate at which energy is absorbed by the planet (Eq. 3–1) and rearrange terms, we obtain an expression for the temperature

$$\text{Temperature} = \sqrt[4]{\frac{\text{Solar flux} \times (1 - \text{albedo})}{4 \times \sigma}} \tag{3-4}$$

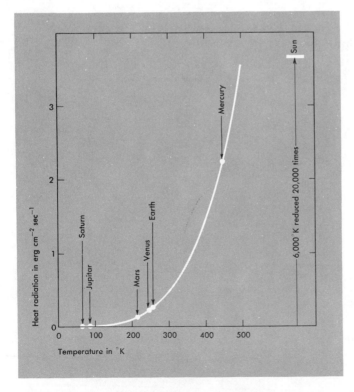

FIGURE 3-3 *The Stefan-Boltzmann law equates the emitted heat radiation to* $5.67 \times 10^{-5} \times (Temperature)^4$. *Effective temperatures of the Sun and planets are shown.*

Atmospheric temperatures

The temperature calculated in this manner is called the *effective temperature* (T_e) of a planet. Note that the radius of the planet has cancelled out of the expression, so the effective temperature depends not on the size of the planet but only on the albedo and the distance of the planet from the Sun. Table 3–1 shows calculated values of the effective temperatures of the planets. Mercury, the planet closest to the Sun, has a very high effective temperature. The outer planets are cold because of their enormous distances from the Sun. Venus and Earth have almost the same effective temperature, even though Earth is farther from the Sun because Venus, with its unbroken cloud cover, has a higher albedo than Earth and absorbs a smaller fraction of the solar radiation incident upon it.

Surface Temperatures

An important distinction must be made between the effective temperatures we have calculated and the temperatures of planetary surfaces. If a planet has a substantial atmosphere, the atmosphere can absorb all heat radiation from the lower surface before the radiation penetrates into outer space. Thus, an instrument in space looking at the planet does not detect radiation from the surface. The radiation it "sees" comes from some level higher in the atmosphere.

The effective temperature is the temperature of this emitting region, and lower levels may have much higher temperatures. On Earth, for example, the average temperature of the surface is 288°K, but the effective temperature is only 253°K. The difference is even more striking on Venus. Ground-based measurements of thermal radio waves emitted by the surface of Venus show that the temperature there is about 700°K, close to the melting point of lead. This surprising result has been confirmed by measurements of atmospheric properties made by the Venera and Mariner spacecraft. But the effective temperature of Venus is only 244°K, close to the effective temperature of Earth. For Mars we expect the average surface temperature to be only a little above the effective temperature, 216°K, because the atmosphere of Mars is thin.

The Greenhouse Effect

The extent to which the temperature of the surface and the lower portions of the atmosphere can differ from the effective temperature depends not only on the mass of the atmosphere but also on its constituents. The important property is the opacity of the gas to electromagnetic radiation and particularly to the infrared radiation emitted by the planet. In Fig. 3–4 we show, at each wavelength, the fraction of the light absorbed from a beam passing directly through the Earth's atmosphere. We also show the shape of the spectrum of incident solar radiation, peaking in the visible because of the high effective temperature of the Sun, and the shape of the thermal emission spectrum of the Earth, peaking far out in the infrared. This is in accordance with our discussion of the Planck spectra in Fig. 1–11. According to *Wien's radiation law*, the wavelength of the peak is inversely proportional to the temperature. Because the effective temperature of the Earth is approximately $\frac{1}{24}$ times the effective temperature of the Sun, the peak wavelength of terrestrial radiation is about 24 times the peak wavelength of solar radiation.

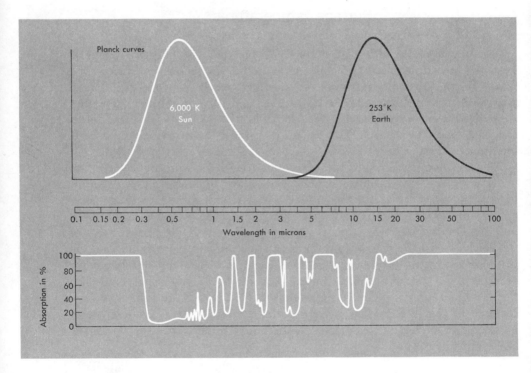

FIGURE 3-4   *At the top are the Planck curves that show the proportion of energy radiated at each wavelength by the Earth and the solar radiation incident on the Earth. There is very little overlap between the curves, which enables us to talk of different atmospheric characteristics for short-wave solar radiation and long-wave heat radiation. At the bottom is the percentage of the radiation at each wavelength that is absorbed in the atmosphere. Absorption is strong on the average for the planetary radiation but weak on the average for the solar radiation. (After R. M. Goody, 1954, Physics of the Stratosphere, Cambridge University Press.)*

We conclude from Fig. 3–4 that the atmosphere is moderately transparent in the visible and that much of the solar radiation can pass right through the atmosphere without being absorbed. On the other hand, minor atmospheric constituents, of which water vapor is the most important, absorb strongly in the infrared so the atmosphere is largely opaque to the planetary heat radiation.

What happens when the atmosphere absorbs radiation emitted from the surface of the planet? The atmosphere cannot steadily accumulate energy or it would become hotter and hotter. Instead, it emits radiation at the same rate as it absorbs. The radiation is reemitted in all directions, and a substantial part of it is intercepted and absorbed by the surface. So the surface of the planet is heated not only by direct sunlight but also by infrared radiation emitted by the atmosphere. For this reason the surface of a planet must radiate away more energy than it receives directly from the Sun, and the surface can have a temperature that exceeds the effective temperature of the planet.

These ideas are given quantitative expression in Fig. 3–5, which shows the mean annual heat budget of the Earth and the atmosphere, derived from meteorological measurements. The left-hand side of the figure shows

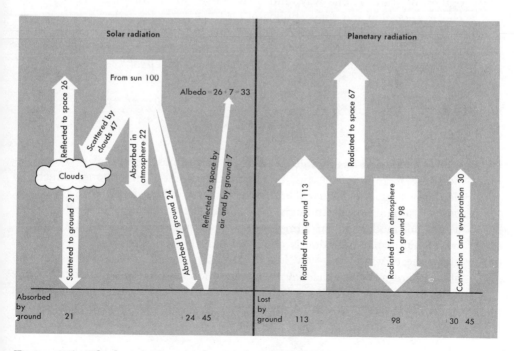

FIGURE 3-5   The heat budget of the Earth and the atmosphere. (After J. London and K. Sasamori, 1971.)

Atmospheric temperatures

what happens to short-wave (visible) solar radiation. All quantities are expressed in units such that the incident solar flux has a magnitude of 100 units. Of this incident flux, about 22 units are absorbed somewhere in the atmosphere and about 33 units are reflected back to space by the ground, the atmosphere, and the clouds. These 33 units represent the Earth's albedo. This energy is lost to the Earth entirely and plays no role in heating either the ground or the atmosphere. The remaining 45 units of the incident radiation are absorbed by the ground. Both the atmosphere and the ground must lose the energy they have absorbed, and they do so by radiating in the long-wave (infrared) region of the spectrum. What happens to this energy is shown on the right of the figure.

The flux of infrared radiation emitted by the ground amounts to 113 of our arbitrary units, two and a half times as much as the flux of solar radiation absorbed by the ground. This extra energy comes, as the figure shows, from the atmosphere. The ground absorbs a long-wave flux from the atmosphere that is equal to 98 units. As a result, the net loss of radiation from the surface in the long-wave region of the spectrum is only 15 units. Since 45 units of solar radiation are absorbed by the ground, we appear to have too much energy going into the ground. The additional ways of removing heat from the ground are shown on the right of Fig. 3–5. Evaporation of water is one way, and heat convection is another. We shall discuss evaporation and convection later on.

This phenomenon, in which the surface temperature of a planet is increased because the atmosphere is translucent to solar radiation but opaque to infrared radiation, is known as the *greenhouse effect.**

Radiative Transfer

We would like to be able to estimate the atmospheric and surface temperature of any planet, with due allowance for the greenhouse effect. To do this we use the theory of *radiative transfer*, which deals with the transport of radiant energy through an absorbing atmosphere. We shall examine a simple model of an atmosphere, in which we assume that short-wavelength solar radiation is absorbed only by the ground and not by the atmosphere, while long-wavelength planetary radiation is absorbed by the atmosphere with an efficiency that is independent of the wavelength of the radiation. Figure 3–4 shows how complete is the separation in wavelength of the solar radiation and the planetary radiation; it is reasonable to assume different optical properties for the two.

*Some writers prefer to avoid this term because the analogy to the domestic greenhouse is not complete, but the term is evocative and rather widely used.

We shall further assume that radiation is the only way of carrying heat from one level of the atmosphere to another. This is called the assumption of *radiative equilibrium*. It has the merit of being the simplest model of heat transfer that has any relevance to our problem, although at a later point we will have to include the effects of convection, condensation, and winds in order to explain observed phenomena. For the present, however, we shall pursue this model to discover its implications.

In order to calculate a radiative equilibrium temperature profile, let us divide the atmosphere into horizontal layers (see Fig. 3–6). The thickness of the layers is adjusted so that radiation emitted in one layer is absorbed in an adjacent layer. The layers must be neither too thick nor too thin. Layers are too thick if radiation is emitted and reabsorbed in the same layer. Layers are too thin if radiation transverses one or more layers before undergoing absorption. Each layer, therefore, is just thick enough to absorb the radiation falling onto it. The mechanism of radiative transfer is one of passing energy from one layer to the next; the radiation emitted from each layer is absorbed by its two nearest neighbors, which in turn emit to their nearest neighbors, and so on (see Fig. 3–6).

The thickness of the layers can generally be expected to increase as the altitude increases because at higher altitudes the density of a thoroughly mixed absorbing atmospheric gas is less, allowing radiation to travel farther before being absorbed. The total number of layers into which an atmosphere can be divided in this manner is called the *optical thickness* of the atmosphere. The optical thickness depends on how much atmosphere there is and also on the efficiency with which atmospheric gases absorb infrared heat radiation.

Let us consider the balance of energy in the topmost layer of the atmosphere (see Fig. 3–6). If the average temperature of the layer is T_1, the layer radiates energy away to space at a rate σT_1^4, and it radiates energy at the same rate downward to the layer below.

First consider the radiation to space. According to our discussion of the effective temperature, on the average the outgoing radiation (σT_1^4) must be equal to the incoming radiation, which in turn is equal to σT_e^4 [see Eq. (3–2)]. From this we conclude that T_e and T_1 are equal.

Now consider the radiation balance of the top layer itself. The only radiation that it can absorb comes from the layer immediately below, at a rate σT_2^4, where T_2 is the average temperature of the second layer from the top. If we now equate the energy radiated by the topmost layer to the energy absorbed, we find

$$\sigma T_1^4 \text{ (upward)} + \sigma T_1^4 \text{ (downward)} = \sigma T_2^4 \text{ (upward)} \qquad (3\text{–}5)$$

This gives us a relationship between the temperatures in the top two layers

$$T_2^4 = 2 T_1^4 = 2 T_e^4 \qquad (3\text{–}6)$$

Atmospheric temperatures

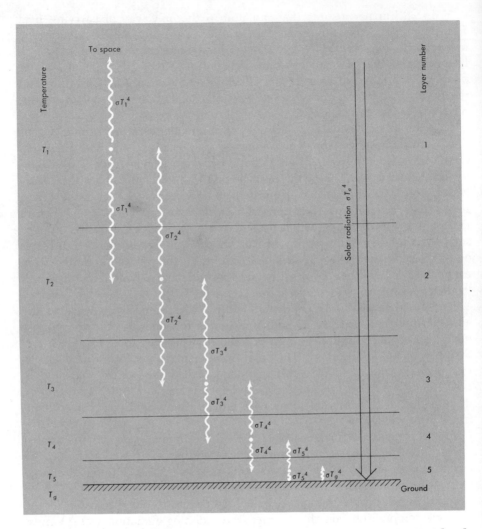

FIGURE 3-6 *Layers of the atmosphere exchanging radiant heat energy with adjacent layers. For this illustration we have chosen an atmosphere that is transparent to solar radiation, but which has an optical thickness of five for planetary radiation.*

In the same way, if we equate the energy radiated by the second layer from the top to the energy absorbed we find

$$\sigma T_2^4 \text{ (upward)} + \sigma T_2^4 \text{ (downward)} = \sigma T_3^4 \text{ (upward)} + \sigma T_1^4 \text{ (downward)} \quad (3\text{-}7)$$

since this layer is heated by layer 1 above as well as by layer 3 below. We find

$$T_3{}^4 = 2\,T_2{}^4 - T_1{}^4 = 3\,T_1{}^4 = 3\,T_e{}^4 \qquad (3\text{-}8)$$

where we have used the fact that $T_2{}^4$ equals $2T_e{}^4$ (Eq. 3–6).

From the energy balance in each successive layer, working down from the top of the atmosphere, we can deduce the temperature of each layer. We have carried the process far enough to see what the answers must be: $T_4{}^4$ is $4T_e{}^4$, $T_5{}^4$ is $5T_e{}^4$, and so on.

Finally, consider what happens at the planet's surface. The energy absorbed is $\sigma T_e{}^4$ from the sun, plus $\sigma T_5{}^4$ $(=5\sigma T_e{}^4)$ from the atmosphere. Energy lost is $\sigma T_g{}^4$ where T_g is the ground temperature. Hence (canceling σ's)

$$T_g{}^4 = T_e{}^4 + 5\,T_e{}^4 = 6\,T_e{}^4 \qquad (3\text{-}9)$$

It is simple to extend the result to an atmosphere of arbitrary optical thickness (and we shall need the result later). We find

$$T_g{}^4 = (1 + \text{optical thickness}) \times T_e{}^4 \qquad (3\text{-}10)$$

Our result is illustrated in Fig. 3–7, where we have plotted, against height, the fourth power of the temperature divided by the temperature in the top layer of the atmosphere, T_1. The figure shows that the temperature decreases steadily with height but approaches a constant value at high altitudes.

In order to plot Fig. 3–7 we have to choose the layers so that each appears equally opaque to planetary radiation. The thickness of the layers depends, in general, on the gas concerned, the optical properties of the gas, and the way in which the gas is distributed in the vertical. To illustrate, let us assume that the absorbing gas is the only gas in the atmosphere and that the density decreases exponentially with a scale height, H.

We have to divide the atmosphere into five slabs, with the same amount of gas in each slab. According to the discussion in Chapter 1, the change in pressure between the top and bottom of a slab is equal to the mass of gas per unit area in the slab times the gravitational acceleration. Thus, the slabs are bounded by surfaces at which the pressure is equal to $\frac{4}{5}$, $\frac{3}{5}$, $\frac{2}{5}$, and $\frac{1}{5}$ times the pressure at ground level.

We have computed the average temperature of each of these layers. The temperatures should now be plotted at the heights corresponding to the middle of each layer, that is, at the levels where the pressure is $9p(0)/10$, $7p(0)/10$, $5p(0)/10$, $3p(0)/10$, and $p(0)/10$, where $p(0)$ is the pressure at the ground.

Now consider the barometric law in the form given in the footnote on p. 9. We can write

$$\frac{z}{H} = -\left\{\frac{1}{\log e}\right\} \times \left\{\log \frac{p(0)}{p(z)}\right\} \qquad (3\text{-}11)$$

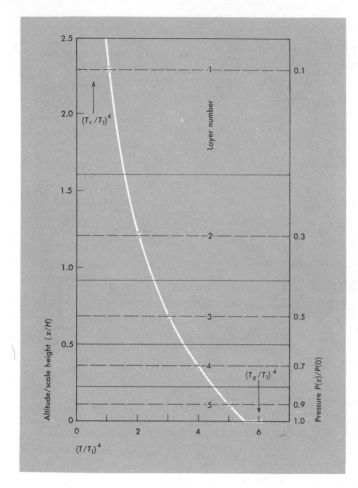

FIGURE 3-7 *The variation of temperature with altitude in an atmosphere of optical thickness 5. The crosses are the results of the calculations in this chapter. The curve is the result of a more detailed calculation. The horizontal solid lines mark the boundaries of the five layers. The broken lines show the mid-points; one-half of the mass of each layer lies on each side of a broken line. The quantities T_1, T_s, and T_g are explained in the text.*

The temperatures we have calculated should be plotted at values of z/H corresponding to $p(z)/p(0)$ equal to $\frac{9}{10}$, $\frac{7}{10}$, $\frac{5}{10}$, $\frac{3}{10}$, and $\frac{1}{10}$. Since $\log e = 0.434$, we find, using logarithms, the values 0.11, 0.36, 0.69, 1.21, and 2.3 for z/H. A similar calculation gives the positions of the surfaces that separate the layers themselves. The layers and midpoints calculated in this way are used in Fig. 3–7. If we are interested in actual heights, then we must multiply z/H by the scale height. For example, if $H = 8.4$ km, as on Earth, the lowest broken line is at a height of 0.92 km and the uppermost is at a height of 19.4 km.

We have now developed the theory of radiative transfer to the point where we can examine, in a quantitative manner, the greenhouse effect on a real planet. An interesting planet in this regard is Venus because Venus has a ground temperature of about 700°K but has an effective temperature of only 243°K. Can the high ground temperature on Venus be a result of the greenhouse effect?

The Greenhouse Effect on Venus

Let us ask how great an optical thickness would be required for the atmosphere of Venus if the greenhouse effect were to provide the high ground temperature. Remember that the optical thickness is the total number of layers into which the atmosphere must be divided so that radiation emitted by each layer is absorbed in the immediately adjacent layers. A massive atmosphere composed of gases that absorb strongly in the infrared region of the spectrum has a large optical thickness. A thin atmosphere has a relatively small optical thickness.

From Eq. (3–10) we can calculate the optical thickness for Venus if we know the effective temperature (243°K) and the ground temperature (700°K) and if we assume that the atmosphere is in radiative equilibrium. We find

$$\text{Optical thickness} = \left\{\frac{700}{243}\right\}^4 - 1 = 68 \tag{3–12}$$

Does the atmosphere of Venus have such a large optical depth, and if it does, is the greenhouse theory the explanation of the high surface temperature?

The first question can be answered to some degree. Carbon dioxide alone cannot give such a large optical depth. The problem lies in the uneven efficiency of absorption in the infrared, which occurs principally in narrow regions of the spectrum (see Fig. 3–4). We require a large optical depth throughout the spectrum including the gaps where absorption is weak. The amount of carbon dioxide known to exist on Venus cannot give the required optical thickness. It is possible that the more transparent gaps are blocked by absorption by other gases that happen to absorb strongly where carbon dioxide absorbs weakly. Water vapor has been proposed as such a supplemental absorber, although we have no direct evidence for its existence in sufficient quantities on Venus. Alternatively, it is possible that the clouds themselves may perform a similar function absorbing strongly throughout the infrared spectrum.

This last suggestion, like all others, has difficulties, and we will not pursue the Venus greenhouse effect further at this stage. We introduced it solely to illustrate principles. It is, after all, quite certain that our model is too simple to explain every feature of the Venus surface temperatures. For example, a radiative equilibrium model predicts on Venus, as on Earth and Mars, high temperatures in the tropics, where sunlight is most intense, and low temperatures in the polar regions. In fact, no significant temperature difference has yet been observed between these locations on Venus. Further discussion must be postponed to Chapter 4, where we examine the effect of winds on atmospheric temperatures.

The Greenhouse Effect on Earth

Let us make a second application of the radiative transfer theory that we have derived, this time to a calculation of the profile of temperature as a function of altitude in the atmosphere of the Earth. Although we know much more about Earth's atmosphere than we know about the atmosphere of Venus, it is not entirely clear how we should divide the atmosphere into layers when we have transparent and opaque spectral regions, as illustrated in Fig. 3–4.

Experience suggests that a reasonable model consists of two layers, with the top layer centered at a height of about 3 km and the bottom layer centered at a height of about 0.5 km. We can derive these heights by the method outlined earlier, if we remember that water vapor is the principal absorbing gas in the Earth's atmosphere. Observations show that the scale height of water vapor is about 2 km.

The temperature of the top layer is equal to the effective temperature. For Earth, this is 253°K (shown by an X in Fig. 3–8 at a height of 3 km). For the bottom layer, the fourth power of the temperature is equal to twice the fourth power of the effective temperature because the bottom layer is the second layer from the top. We find that the temperature of the bottom layer is 297°K (shown by an X in Fig. 3–8 at a height of 0.5 km).

Since these two data are not enough to draw a profile, we will compute one further point, namely, the lower limit to the temperature at very great heights. We hinted at the existence of such a limit in Fig. 3–7. A detailed computation shows that the temperature never drops below $(T/T_e)^4 = 1/2$ and tends to this limit as the height increases.

To show that this is the appropriate limit, consider Fig. 3–9, which shows a very thin layer high in the atmosphere, and hence, far above the average height of layer 1. The *opacity* of this layer is ε and this we assume to be a very small quantity.

The opacity specifies the capacity of a layer to absorb radiation. Hence, from the upwelling radiation from layer 1, which equals σT_1^4, a fraction $\varepsilon \sigma T_1^4$ is absorbed in the thin layer. This is indicated in Fig. 3–9 by a beam of flux σT_1^4 incident from below and a beam of flux $\sigma T_1^4(1-\varepsilon)$ leaving to space.

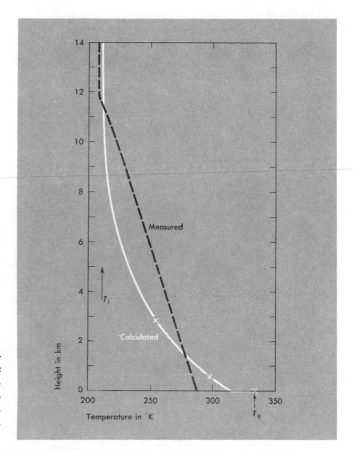

FIGURE 3-8 *The temperature profile in the Earth's atmosphere calculated under the assumption of radiative equilibrium compared with average measured temperatures.*

The temperature of the thin layer is T_s, called the *skin temperature*. If the layer were opaque, it would emit σT_s^4 in both directions, up and down. It is not opaque, however, because it is very thin, and we must appeal to one of *Kirchoff's radiation laws*, which tells us that for translucent bodies *emissivity* and *opacity* are equal; both are therefore equal to ε. The total emission by the layer is therefore $2\varepsilon\sigma T_s^4$, one half upward and one half downward.

If the layer is in radiative equilibrium, energy absorbed equals energy emitted or

$$\varepsilon\sigma T_1^4 = 2\varepsilon\sigma T_s^4 \tag{3-13}$$

Upon rearranging, we find

$$\left(\frac{T_s}{T_1}\right)^4 = \frac{1}{2} \tag{3-14}$$

Atmospheric temperatures

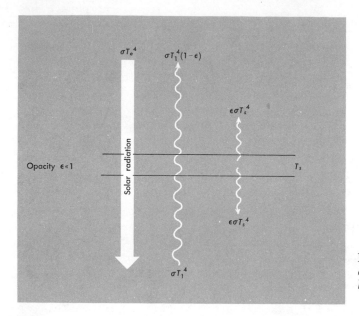

FIGURE 3-9 *Radiative equilibrium in a very thin layer at the outer limit of the atmosphere.*

which is the result we set out to derive.* For Earth, with $T_1 = T_e = 253°K$, the skin temperature is $T_s = 212°K$.

Our theoretical model will be complete once we have calculated the ground temperature. The fourth power of the ground temperature is equal to the fourth power of the effective temperature added to the fourth power of the temperature of the bottom layer of atmosphere. We find a value for the ground temperature of 333°K, shown by an arrow at the bottom of Fig. 3–8.

We can now compare the theoretical temperatures we have just calculated with average temperatures measured in the real atmosphere. Measured temperatures are shown by the broken line in Fig. 3–8. We see that the theoretical model is quite successful at altitudes above 10 km but that there are substantial deviations throughout the troposphere. Our theory is inadequate because radiation is not the only process that carries heat upward from the ground and from the lower levels of the troposphere. Another process tending to hold down the temperature at the ground and to increase the temperature of the upper troposphere is known as *convection.* Under certain circumstances convection can carry heat away from the ground more effectively than can radiation.

---

*Higher temperatures result if the upper levels of an atmosphere absorb solar radiation directly, as is the case in the terrestrial ozone layer. This topic is developed in the Appendix.

Heat Transport by Convection

The mechanism of convection is familiar to all of us. It occurs because air expands when it is heated, causing the density to decrease (see our discussion of the ideal gas law in Chapter 1). If we have abnormally hot air near the ground, it will be lighter than its surroundings and, like a balloon, will attempt to rise. Similarly, if there is abnormally cool air at high levels, it may be dense compared with its surroundings and may tend to sink. Low level warm air will replace high level cold air and vice versa, giving a net upward transport of heat. We must now determine the conditions under which this process of convection takes place and how it affects the temperature structure of an atmosphere.

Our first problem is that the temperature of a rising parcel of air tends to fall as the parcel rises because the atmospheric pressure decreases with altitude at a rate given by the barometric law (see Chapter 1). Thus, a rising parcel of air moves from a region of higher pressure to a region of lower pressure and expands as it does so. The temperature of the air decreases as the air expands, just as does the temperature of air released from a tire. The expanding parcel pushes back the air pressing on its boundaries and, therefore, does work upon the surroundings. Work is a form of energy, and conservation of energy requires that some other form of energy in the parcel must decrease. The only possibility for dry air is the internal thermal energy of the molecules. Consequently the temperature must drop as the parcel rises.

To an acceptable degree of approximation we may regard this decrease of temperature with height as independent of the state of the surrounding atmosphere. In Fig. 3–10, the solid line AB represents the variation of temperature with height in a parcel of air rising through the atmosphere. The rate of decrease of temperature with height corresponding to AB is called the *adiabatic lapse rate* (deg cm^{-1}). We shall calculate its value below. Let us accept its existence for now and consider what happens in an atmosphere when temperature varies at a different rate. If the temperature of our rising parcel of air decreases more rapidly than the temperature of the surrounding atmosphere, the parcel will always be colder and denser than the atmosphere around it (see Fig. 3–10). Under these circumstances its excess weight will tend to drag it back down to the atmospheric level from whence it came, and convection will not occur. The atmosphere is then said to be *stable*.

On the other hand, if the temperature of the atmosphere decreases with altitude more rapidly than the temperature in a rising parcel of air, we

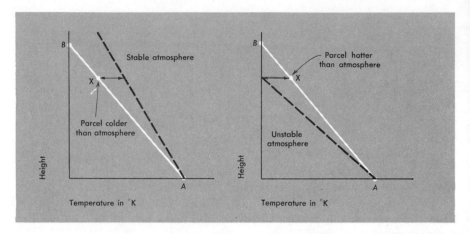

FIGURE 3-10 *The temperature profile in a stable atmosphere is shown by the dashed line in the left-hand diagram. The dashed line in the right-hand diagram shows the temperature profile in an unstable atmosphere. The solid line in each diagram has a negative slope equal to the adiabatic lapse rate; it is approximately the temperature of a parcel of air rising rapidly through the atmosphere. The parcel starts at the bottom point A in each case, where it has the same temperature as the surrounding atmosphere and rises along the path AB. When the parcel on the left reaches point X, it is colder than the surrounding atmosphere and therefore denser. It tends to sink back to the level where it started. When the parcel on the right reaches point X, it is hotter and therefore less dense than the surrounding atmosphere. Buoyancy causes it to rise still further.*

have the situation we envisaged when we began the discussion of convection. After having risen a short distance, the parcel of air will have a higher temperature than the surrounding atmosphere in spite of its temperature having fallen as a result of expansion. Because it is hotter, the density of the air in the parcel will be lower than the density of the atmosphere, and buoyant forces will accelerate its upward motion. Thus, the atmosphere is in a condition whereby it can change spontaneously, without outside intervention, like a pen balanced precariously on its tip. This circumstance we term *unstable*.

We conclude that convection can only occur spontaneously if the temperature of the atmosphere decreases quickly enough with altitude. An isothermal atmosphere, for example, is stable against convection. Although the temperature in the troposphere generally decreases with increasing altitude, there are times when the air near the ground is colder than the air above, giving a temperature that increases with altitude instead. This meteorological phenomenon is called an *inversion*. During an inversion, convection is strongly inhibited and very little mixing occurs from one level of the atmosphere to another. As a result, pollutants accumulate in the air

Atmospheric temperatures

near the ground instead of being dispersed to higher levels. It is at times of inversion that air pollution is most severe.

We have already pointed out that rising parcels of air follow temperature curves very similar to each other (curve AB in Fig. 3–10). We can evaluate the slope of this line relatively easily when the surrounding atmosphere also has temperatures lying on the curve AB: a parcel of air then can rise and fall in a neutral fashion without exchanging heat with the surroundings. When no heat is exchanged, we refer to the change as *adiabatic*. The calculation in the Appendix shows that the temperature falls at a rate

Adiabatic lapse rate $(\deg \text{cm}^{-1}) =$

$$\frac{\text{Acceleration of gravity } (\text{cm sec}^{-2})}{\text{Specific heat of air at constant pressure } (\text{erg deg}^{-1} \text{ gm}^{-1})} \quad (3\text{–}15)$$

Values of the adiabatic lapse rate of different planets are given in Table 3–2.

Convection is an efficient means of transporting heat in a gas. Both experience and theory show that the rate of decrease of temperature with height in a planetary atmosphere cannot exceed the adiabatic lapse rate by a substantial amount except very close to the ground, where convection is inhibited. Imagine an atmosphere in which the temperature lapse rate is initially less than the adiabatic lapse rate and imagine that the lower levels are heated while the upper levels are cooled, so that the lapse rate increases with time. Initially the atmosphere is stable against convection and no motion occurs. As soon as the lapse rate exceeds the adiabatic lapse rate, however, the atmosphere becomes unstable against convection. Convection carries heat from the hot lower levels of the atmosphere to the cold upper levels, inhibiting further increase of the lapse rate.

We can now decide whether or not the theoretical profile of temperature in a planetary atmosphere is stable against convection. When the predicted rate of decrease of temperature with altitude is less than the adiabatic lapse rate, the atmosphere is stable, and convection will not occur to modify the temperature profile. But whenever the predicted lapse rate

Table 3–2

Adiabatic Lapse Rates

Planet	Gas	Gravitational Acceleration (cm sec^{-2})	Specific Heat (erg gm^{-1}deg^{-1})	Adiabatic Lapse Rate (deg km^{-1})
Venus	CO_2	888	8.3×10^6	10.7
Earth	N_2, O_2	981	1.0×10^7	9.8
Mars	CO_2	373	8.3×10^6	4.5
Jupiter	H_2	2620	1.3×10^8	20.2

Atmospheric temperatures

exceeds the adiabatic lapse rate, convection will occur, and the temperature profile will be modified so that temperature decreases at about the adiabatic rate.

Our examination of the temperature profile in an atmosphere transporting heat only by radiation showed that the lapse rate is small at high altitudes, where there is little absorbing gas, but increases steadily as the altitude decreases; close to the ground it becomes very steep indeed, as indicated by the horizontal portions of Fig. 3–7 and Fig. 3–8. We can, therefore, conclude that an atmosphere will be stable against convection at higher altitudes but may become unstable at the lower altitudes where there is enough gas to absorb infrared radiation strongly.

These considerations and their effect on the temperature profile for Earth, are shown in Fig. 3–8. At altitudes above 12 km, the temperature is not influenced by convection, and the measured temperature profile follows the radiative equilibrium solution. At lower altitudes, convection takes over from radiation as the most important heat transport process and the temperature profile becomes a straight line.* The net effect of convection, as the figure shows, is to reduce the ground temperature by about 60°K.

The theory we have described, including heat transport by convection as well as by infrared radiation, is reasonably successful at explaining the average temperature profile in the lower few tens of kilometers of the Earth's atmosphere. The layer at the bottom of Fig. 3–8, where convection is important and the temperature decreases steadily with increasing height, corresponds to the troposphere shown in Fig. 3–1. The overlying layer, where convection is not important and the temperature is very nearly constant, corresponds to the lower stratosphere. The temperature profile in the upper stratosphere and mesosphere can also be understood in terms of radiative transfer and convection but, as we have already noted, it is necessary to take into account the heat source provided, at heights around 50 km, by the absorption of solar ultraviolet radiation by ozone (see Appendix).

The radiative-convective theory is not applicable at heights above about 80 km in the Earth's atmosphere where the air is too thin to interact strongly with radiation. We shall discuss this high altitude region (the thermosphere) later in this chapter. Before we do so, let us consider the use of the radiative-convective theory to predict temperatures in the lower atmospheres of some of the other planets.

First consider Venus. If we apply the radiative-convective theory, we find that beneath the cloud tops the temperature should vary at the adiabatic lapse rate. As far as we can tell from space probe measurements,

---

*The average lapse rate in the troposphere is 6.5°K km⁻¹, which is smaller than the adiabatic lapse rate given in Table 3–2. The reasons for this difference are known. We have correctly described some essential physical processes, but not all of them. Two other factors are large-scale planetary motions (discussed in Chapter 4) and the condensation of water (discussed in Chapter 5).

this prediction is verified. Introduction of convection, however, does not eliminate the difficulty that the temperature is the same at the equator as at the poles. Our theory still predicts that $\sigma T_e{}^4$ is equal to the absorbed flux of solar radiation; although this is small at the poles and large at the equator, no variation of T_e is, in fact, observed. Thus, while these ideas are helpful in picturing factors involved in the high ground temperatures and the adiabatic lower atmosphere, they are definitely over-simplified.

The Martian Troposphere

For Mars, let us first consider an average over day and night conditions in the Martian tropics. The theory predicts an average temperature profile that resembles the terrestrial profile shown in Fig. 3–8, except that temperatures are lower on Mars because Mars is farther from the Sun. In the bottom 15 km of the tropical Martian atmosphere the temperature decreases at the adiabatic lapse rate from a value of about 230°K near the ground. Above 15 km the temperature is nearly constant and has a value of about 155°K. The region of temperature decrease corresponds to the troposphere on Earth; convection plays a dominant role in maintaining the average temperature profile. The overlying region corresponds to the lower stratosphere on Earth, where heat is transported mainly by radiation. We have already mentioned that we do not expect to find a temperature maximum at intermediate heights in the Martian atmosphere, corresponding to the terrestrial stratopause (Fig. 3–1), because Mars does not have enough oxygen in its atmosphere to produce ozone.

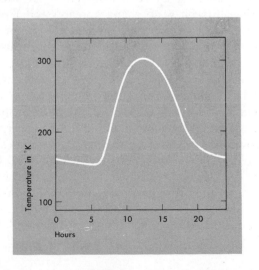

FIGURE 3-11 *Diurnal variation of the ground temperature at the equator of Mars during an equinox. The temperature during the day has been measured. The night-time values are theoretical. (After P. Gierasch and R. M. Goody, 1968.)*

Atmospheric temperatures

The Martian troposphere differs from the terrestrial troposphere in one important respect. On Mars there is a large diurnal variation of the temperature. The temperature of the ground at the equator of Mars is shown in Fig. 3–11. At noon it is about 300°K, not very different from temperatures in the Earth's tropics. At night, on the other hand, the Martian temperature drops to a frigid 160°K, much colder than any place on the surface of the Earth. This large diurnal oscillation is in part a result of the absence of oceans on Mars. Our oceans are partially transparent to sunlight and are also in a continual state of motion. Thus, a substantial layer at the top of the ocean absorbs heat during the day and loses heat during the night. A thick layer of water can hold a large amount of heat, so very little temperature change results from the diurnal cycle of heating and cooling.

On Mars, however, in the absence of oceans, all the daytime heating and night-time cooling must take place right at the surface of the ground. The thermal conductivity of the Martian surface is believed to be small, with only a thin layer taking part in the diurnal oscillation. Since a thin layer of dry surface material can hold very little heat, large temperature changes result from the diurnal cycle of heating and cooling. (We can observe this phenomenon on Earth, where diurnal surface temperature changes are much larger in dry desert areas a long way from the sea than they are on or near the oceans or on land covered with vegetation. Observations on Mars indicate that the surface does indeed resemble light, dry desert sand.)

The large variation in the ground temperature on Mars produces a correspondingly large diurnal variation in the flux of infrared radiation. Since this flux is mainly responsible for heating the lower atmosphere, we may expect a corresponding variation in atmospheric temperature. Because of its composition, the Martian atmosphere is much more sensitive than the terrestrial atmosphere to changes in infrared flux.

On Mars the atmosphere contains a large proportion of carbon dioxide molecules (see Chapter 1), which absorb infrared radiation. On Earth, on the other hand, the most important absorbers are water vapor molecules, which constitute a very small fraction of the total atmosphere. Heat absorbed by a molecule of water vapor on Earth must be shared amongst hundreds or thousands of oxygen and nitrogen molecules which themselves do not absorb heat radiation. On Mars, however, each carbon dioxide molecule absorbing or emitting thermal radiation can use the heat to change its own thermal energy. The response of the atmospheric temperature to changes in radiation is, therefore, far more rapid for Mars than for Earth; so the large diurnal temperature oscillation on Mars occurs not only at the ground but is also transmitted by radiation into the atmosphere up to a height of several kilometers. Theoretical profiles of Martian temperatures at different times of day are shown in Fig. 3–12. On Earth, by way of con-

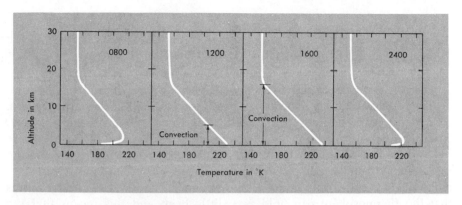

FIGURE 3-12 *Theoretical temperature profiles in the Martian atmosphere at different times of the day (after P. Gierasch and R. M. Goody, 1968.) At 0800 hours the ground is colder than the overlying atmosphere, and there is no convective region. Convection starts before 1200 hours, is stronger at 1600 hours, but dies away at night. At 2400 hours the ground is colder than the atmosphere once again. Ground temperatures are given in Fig. 3-11. There is a shallow layer near the ground in which temperature changes very rapidly with height.*

trast, diurnal temperature oscillations are small except in the few tens of meters closest to the ground.

The diurnal variation of temperature in the lowest levels of Mars' atmosphere causes a corresponding variation in the occurrence of convection. During much of the night, as Fig. 3–12 shows, the ground is colder than the lower atmosphere, and the temperature actually increases with height in the first one or two kilometers (there is an inversion). A temperature profile such as this is stable against convection. Even at higher altitudes, where the temperature does decrease with height, it does so at a rate less than the adiabatic lapse rate and the atmosphere remains stable. The result is that there may be no region of convective instability in the Martian atmosphere at night.

After the sun rises, however, theory indicates that the ground temperature increases rapidly to values considerably in excess of the atmospheric temperature. A region of convective instability develops, close to the ground at first, but extending higher and higher as the atmosphere warms up. By the end of the day Mars has a well-developed convective layer extending as high as 15 km. Soon after nightfall the ground and the lower atmosphere cool down again, and convection ceases. The predicted diurnal variation in the thickness of the convective layer is, as far as we know, peculiar to Mars.

Temperature in the Thermosphere

There is a region of the Earth's atmosphere where diurnal temperature changes are very large indeed, as they are in the Martian troposphere. This region is the thermosphere, at altitudes above about 100 km (see Fig. 3–1). The density of the gas at these heights is low. At 120 km there are fewer than 10^{12} molecules per cubic centimeter, and at 1000 km there are fewer than 10^6 molecules per cubic centimeter. (There are more than 10^{19} molecules per cubic centimeter at the ground.) The low gas density means that the thermosphere can hold very little heat, and this means that the temperature of the thermosphere can respond rapidly to changes in the amount of solar radiation absorbed at these levels. The result is a substantial diurnal variation in thermospheric temperature, with afternoon temperatures exceeding early morning temperatures by 300°K or more.

The gas densities in the thermosphere are low, but the temperatures are generally high. The mean temperature of the neutral gas increases from 350°K at 120 km to about 1000°K at 250 km. Above this height the temperature is very nearly constant. One might think that these high temperatures are evidence of a large source of energy, but such is not the case. The total rate of heating in the thermosphere is only about one-millionth of the rate at which the ground is heated by visible radiation from the sun. Instead, the high temperatures reflect the inefficiency of mechanisms for removing heat from this region of the atmosphere. For although in the lower atmosphere heat is transported efficiently by infrared radiation emitted by water vapor, carbon dioxide, and ozone, in the thermosphere infrared radiation is relatively unimportant.

Temperature, as normally defined, has to do with the energy of translational motion of the molecules. Absorption and emission of infrared radiation, on the other hand, are concerned with the increase or decrease of the rotational and vibrational energy of molecules. If collisions are rapid, these three types of energy—translation, rotation, and vibration—readily interchange. An excess translational energy leads to excess vibrational energy that can then be emitted as heat radiation. In this manner heat radiation affects the temperature of the gas as described by the Stefan-Boltzmann law.

If the pressure is very low, however, and collisions are infrequent, the different forms of energy are not rapidly shuffled. High temperature leads to high velocities and high translational energy but not necessarily to the high vibrational and rotational energies that affect heat radiation. This problem is illustrated schematically in Fig. 3–13. Up until now we have assumed the energy transfer link (bottom of the diagram) to be rapid; thus,

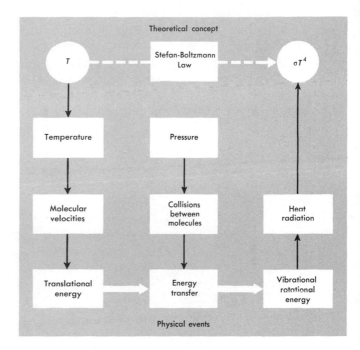

FIGURE 3-13 *The relationship between the theoretical concept of heat radiation (Stefan-Boltzmann law) and the physical chain of events. If pressure is low, the rate of energy transfer is low and high temperature no longer implies high heat radiation or vice versa.*

temperature T led naturally to heat radiation σT^4. But if the link is slow because of infrequent collisions, heat radiation has little connection with temperature. This is the situation in the thermosphere. If we want to lose or gain heat in order to change the temperature, we must find another way besides thermal radiation.

The other form of heat transfer that we have discussed so far is convection. The thermosphere, however, is strongly stable against convection because temperature increases with height, and we anticipate almost no heat transfer from this cause. The only remaining possibility is the transfer of heat by direct collisions between molecules from high temperature regions and molecules from low temperature regions. This is the mechanism of thermal conduction. Since conduction in a gas is a relatively inefficient means of transporting heat, large temperature differences are needed to conduct away the small amount of solar heat that is absorbed in the thermosphere. The small thermospheric heat source is provided by solar ultraviolet radiation with wavelengths shorter than 1000 Å, which is absorbed at heights of about 200 km (see Fig. 2–4). This is the radiation that photoionizes the atmosphere and produces the ionosphere, a subject we discussed in Chapter 2.

The heat must be conducted all the way down to about 100 km, where collisions are sufficiently frequent to allow infrared emission to remove the heat. This is the reason for the substantial increase in temperature between

100 km and 200 km (see Fig. 3–1) and for the high temperature in the thermosphere.

Satellite measurements over a period of years have revealed wide fluctuations in the temperature of the thermosphere; daytime temperatures have climbed as high as 1800°K and have fallen as low as 900°K. These excursions are caused by changes in the flux of ultraviolet radiation from the Sun.

4

Winds of global scale

Atmospheres in Motion

Winds are an everyday experience; their relationship to the weather and their consequent importance in matters economic, social, and recreational requires no emphasis. The objective of this chapter is to explain some characteristics of wind systems, particularly those that concern the planet as a whole.

A familiar characteristic of winds is that they change from hour to hour, from day to day, and from place to place; it is just this variability that concerns the meteorological forecaster. Rather than concentrate on the variability, however, we shall devote most of our attention to the mean winds, which describe the motion of the atmosphere averaged over many days or weeks and over areas as large as continents.

If we concentrate attention upon the average global winds, we have to assume that the variability associated with weather systems, storms, and fronts is, in some way, a matter of secondary importance that can be treated after the broad structure of the wind systems has been understood. We must emphasize at the outset that this cannot be true in any exact sense, for average and variable winds are closely connected; we shall show that they affect each other in an intimate way. An alternative approach is to examine the detailed behavior of weather elements and from this attempt to build a picture of the global winds. This approach, however, is so intricate that it makes it difficult to gain a clear picture of the forces acting on the atmosphere and the response of the atmosphere to these forces.

For the global average winds, we shall use the name *general circulation*. For more than two centuries the general circulation has been regarded as a distinct phenomenon that can be understood as the result of a direct causal relationship between forces and motions. In recent decades increasing emphasis has been placed on the mutual interactions between the average and variable components of the winds. This has led to much more complicated models, but the concept of the general circulation is still considered valuable.

For planets other than Earth the emphasis has to be on the global wind systems, since we are never likely to have many observations of the detailed and variable elements. Although we cannot observe them we are not, however, completely at a loss. Since computers can handle models of the Earth's atmosphere containing most of the detail of a conventional weather map, we may perhaps be able to predict the action of weather systems on other planets, at least in a broad sense. Furthermore, our discussion of global winds will lead us to the point of understanding some features of the genesis of weather systems. If our ideas prove correct on the Earth, we may anticipate success when we apply them in an appropriate form to other planets.

General Circulation
of the Terrestrial Atmosphere

We must now examine in more detail the concept of averaging meteorological observations in order to eliminate weather fluctuations and to isolate the global motions.

Figure 4–1 shows a ground level weather map for a single day over the entire Northern Hemisphere. We are all familiar with maps that show lines of constant pressure or isobars, expressed in terms of millibars (one thousand times the C.G.S. unit of pressure). Later we shall show that the isobars are, under some circumstances, associated with the flow pattern of the wind, and that the wind tends to flow along isobars. Thus the picture presented in Fig. 4–1 is one of a disorderly field both of pressure and of wind.

Despite the disorderly appearance in Fig. 4–1, if we measure the wind where it is little affected by local structures and topography, there tends

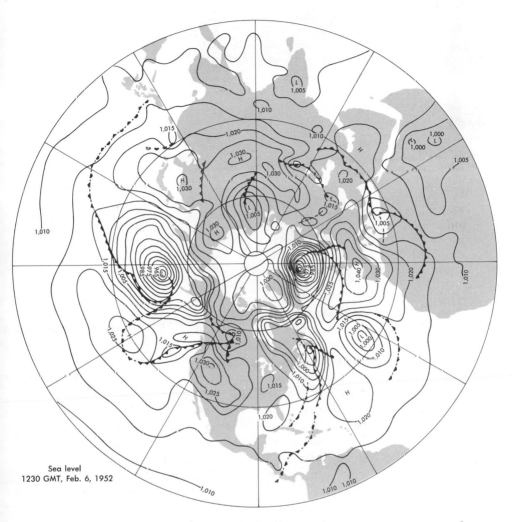

FIGURE 4-1 *The weather map of the Northern Hemisphere. (From General Meteorology by H. R. Byers. Copyright 1959, McGraw-Hill. Used by permission of McGraw-Hill Book Company.)*

to be a prevalent wind direction and a typical average wind strength. For example, airlines working between New York and Europe allow significantly less elapsed time for the flight to Europe. There is a rather dependable *westerly* wind (a wind from the west) at aircraft altitudes over the Atlantic Ocean.

Before discussing these global winds let us briefly consider how we average wind vectors, which describe both the direction and magnitude of the wind. In Fig. 4–2 the distances travelled by the wind on four successive

Winds of global scale

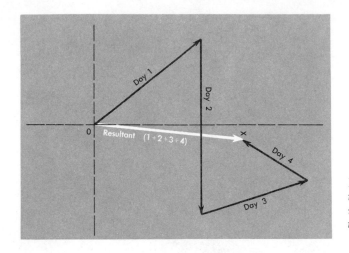

FIGURE 4-2 *Vector addition of winds. The daily mean wind is in the direction OX but is one-quarter the length.*

days are represented by arrows joined beginning to end. Starting from 0 we arrive at X after four days; hence the arrow OX represents, in magnitude and direction, the steady wind that would give the same displacement in one day; one-quarter of this is the *daily average wind.*

If the average wind is to have significance, it must be relatively constant from one year to another. Figure 4–3 shows the average wind for each year in the 1960's at the Blue Hill Observatory in Milton, Massachusetts. All of the mean winds have a magnitude between 14 and 15 mph (6.2 to 6.7 m sec^{-1}) and they are almost all in the segment W to WNW (note that the points of the compass are reversed in the figure because meteorological usage names a wind after the direction from which it comes). We may conclude that a mean wind of 6.5 m sec^{-1} about 10° N of W is an important feature of the global wind system in Eastern Massachusetts.

Now suppose that instead of plotting daily maps of the weather, we were to evaluate the average pressure or wind over one particular month and plot these averaged data on a map. A typical result is shown in Fig. 4–4. The map is for a height of about 5.5 km, well away from the ground, so that surface disturbances are minimized. Compared to Fig. 4–1, the map is far more regular in appearance. There is now a marked tendency for high pressures to occur in tropical regions and low pressures in the polar region. The isobars do not deviate greatly from circles, and the deviations that do occur appear to be associated with large land masses.

Let us agree to remove the deviations related to continents, at least for a preliminary discussion. This we may do by averaging quantities in which

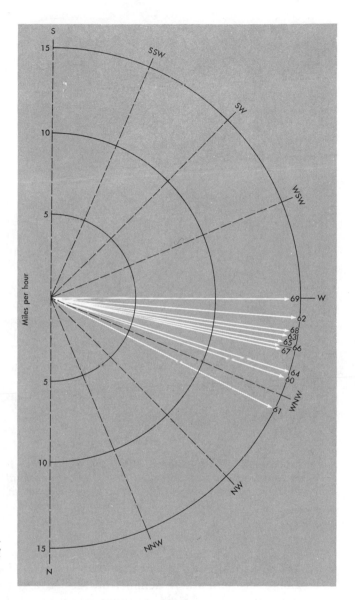

FIGURE 4-3 *Annual average winds at the Blue Hill Observatory in Milton, Massachusetts.*

we are interested, such as pressure, wind, or temperature, around circles of latitude. This procedure leads to quantities that depend only on latitude and on height; they can, therefore, be plotted on a single sheet of graph paper.

Winds of global scale

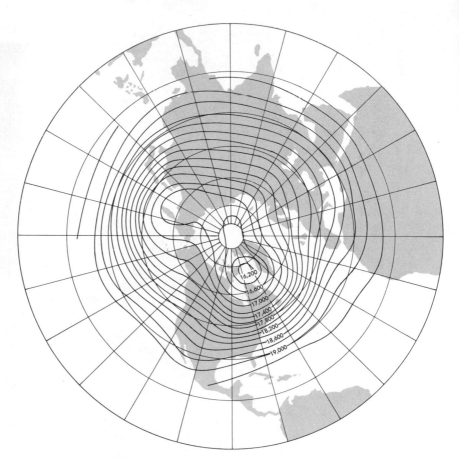

FIGURE 4-4 *Mean contours of the 500 mb surface for the Northern Hemisphere in January. The numbers give the height, in feet, at which the pressure is 500 mb. The map is similar to the pressure map at an altitude of about 5.5 km.)*
(From A. Miller, 1971, Meteorology, Charles E. Merrill Publishing Co.)

Figure 4–5 shows the average east-west wind components over the whole globe, for the northern summer on the left and the northern winter on the right. Instead of height, pressure has been used as a vertical coordinate. Since the two are directly related by the barometric law (p. 8), we may make this transformation at our convenience. The pressure decreases outward along a radius. The resulting impression is as if a slice had been taken through the planet and atmosphere together.

Figure 4–5 shows only east-west components of the wind. These are the more interesting because average winds blow much more in east-west directions than in north-south directions (see Fig. 4–3, for example). Some

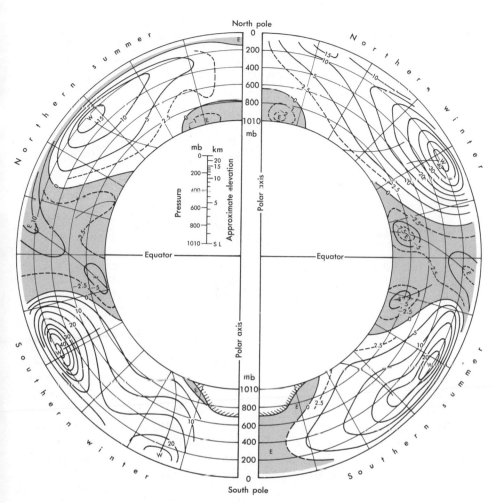

FIGURE 4-5 *Mean zonal wind (east-west) averaged over latitude circles for northern summer (left) and northern winter (right). Winds are in m sec⁻¹. Easterly winds are shaded. The radial scale is pressure rather than height. (From Y. Mintz, 1954, Bull. American Meteorological Soc.)*

more light is shed by Fig. 4–6 which, unlike Fig. 4–4, is highly schematic and should only be used qualitatively.

Figure 4–6 indicates the main surface wind systems with their nautical names. The north-south components of the velocity are greatly exaggerated in this figure; in fact they are small compared to the east-west components, but they are not negligible. We can represent these north-south components by means of the closed circulation cells shown round the perimeter of Fig.

Winds of global scale

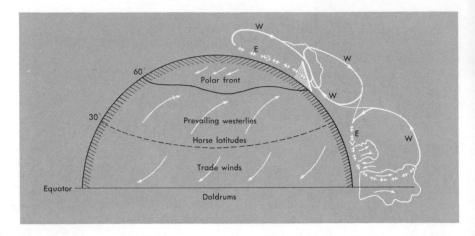

FIGURE 4-6 *A schematic view of the mean meridional winds (north-south) according to C.-G. Rossby (1940). The zonal winds are indicated by the direction of the surface arrows and the letters E for East and W for West. The magnitudes of the north-south components of the surface winds are exaggerated.*

4–6. Thus, for example, one cell suggests a surface wind from north to south between 30°N and the equator. We have to combine this with strong surface easterly winds as shown at low latitudes in Fig. 4–5. The result is a north-easterly wind known as the *northeast trade wind.*

We should emphasize immediately that the existence of the cells shown in Fig. 4–6 is debatable. This particular three-cell structure was first suggested by Ferrel in 1856. The cell nearest to the equator is the easiest to explain, as we shall show, and is called the *Hadley cell.* Its presence has been established directly from averaged meteorological data. The mid-latitude cell, in the reverse direction to the Hadley cell, is partially established, but some people question its reality. The third, high latitude cell is even more questionable.

Weather and Turbulence

We have separated a general circulation or global system of winds from the variable weather elements by means of averaging, but we must be careful to understand what may be the physical consequence of so doing, for we have already remarked that the two types of wind are related. As a crude analogy, let us consider the simple laboratory experiment originally performed by Reynolds.

Winds of global scale

In this experiment water flows along a horizontal glass tube (from left to right in Fig. 4–7.) A hypodermic needle is placed in the stream, and a slow flow of ink is forced through it, giving a visible indication of the motion of the water. If the water flows slowly enough, we find the situation shown in Fig. 4-7(a). The flow is smooth, and the ink remains in a narrow stream parallel to the axis of the tube. The flow velocity at various distances from the center is steady and parallel to the axis, as shown on the right of the figure. This smooth flow is well understood and can be calculated theoretically from knowledge of the physical properties of water.

As the velocity of flow is gradually increased, the flow remains smooth until a certain critical velocity is reached. Suddenly the whole flow becomes unsteady. It oscillates to and fro across the tube, as illustrated in Fig. 4–7(b). The flow is now referred to as *turbulent*, and it obviously differs greatly from the previous smooth flow.

In an analogous fashion to our treatment of the general circulation, we note that the flow at each point fluctuates erratically; we can, however, concentrate our attention on the mean flow averaged over a long time interval. Since the water is flowing from left to right, it follows that the average velocity of the water must be from left to right, parallel to the axis. We may, therefore, draw the average flow velocity diagram shown on the right in Fig. 4–7(b). This average flow is our analogy to the general circulation. The turbulence in our experiment is the analogue of the weather systems in the atmosphere.

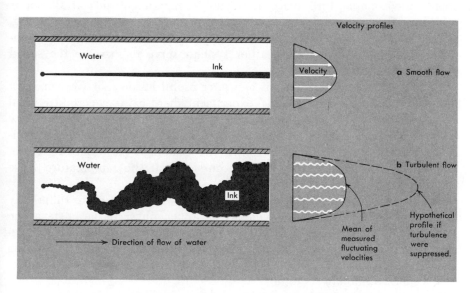

FIGURE 4-7   *Reynold's experiment with turbulent flow.*

Winds of global scale

We can now ask to what extent it is reasonable to develop theories in terms of the average flow when the fluid is, in fact, turbulent. Suppose we had calculated the velocity distribution neglecting the turbulence complently, as we did in the case of smooth flow, using the same driving force as for the turbulent flow. The result of this calculation is shown by the broken line on the right in Fig. 4–7(b).

From this comparison we learn some important lessons. The smooth and turbulent flows correspond to the extent that the average flow is still in the same direction. Moreover, in both cases there is a maximum value in the center of the tube. On the other hand, we calculate velocities that are much too high if we ignore the existence of the turbulence.

We may reasonably draw two conclusions about the general circulation. First, erratic variations with time and place associated with weather will affect the global flow of the atmosphere; a complete theory must include an account of this large-scale "turbulence." Second, if we consider only nonturbulent models, we may reach conclusions that are correct in some essential respects. We may, for example, expect to obtain a qualitative account of the response of the fluid to imposed forces; the pressure force driving the water down the tube in Reynolds' experiment is one such force (see next section).

We must be careful in our thinking about atmospheric motions not to be misled by the analogy to Reynolds' experiment. There are important differences between the effect of turbulence on the mean flow of the water in a tube and the effect of weather on the general circulation of the atmosphere. For example, turbulence in Reynolds' experiment acts at all points to reduce the average flow velocity, as the diagram on the right in Fig. 4–7(b) shows. In the atmosphere, on the other hand, there are circumstances in which fluctuating weather systems serve to increase the speed of the average wind.

In spite of this caveat, there is a further useful lesson that we can learn from the laboratory. In Reynolds' experiment there exists a certain critical velocity below which smooth flow occurs and above which turbulent flow becomes the normal mode of behavior of the fluid. We encountered a similar critical transition in Chapter 3 when we examined the onset of convection in the atmosphere. Convection occurs only if the temperature lapse rate exceeds a critical value.

We will again meet the idea that a fluid may behave quite differently depending on the value of a velocity or other physical parameter. The smooth flow is sometimes called a *stable flow* and the turbulent, changeable flow is called an *unstable flow*. The critical condition for a transition from stable to unstable behavior (in this case the magnitude of the velocity) is called a *stability condition*.

Pressure Forces

All fluids, whether gases or liquids, possess the property of viscosity, a dissipative force that opposes motion. For the present, we do not need to know more about viscous forces than that they are always at work in a fluid if it is in motion relative to itself or to boundaries. Thus, according to Newton's laws of motion, the fluid must be driven by other forces in order to maintain a steady state of motion. If not, the motion will gradually decrease to zero as a result of viscosity. Since we observe that atmospheres are in continuous motion, we have to identify and discuss the driving forces responsible for keeping them in that state. The most fundamental of these forces is the *pressure force*.

In Chapter 1 we defined the pressure at a given level in a fluid as the force per unit area needed to balance the downward force of gravity on all of the fluid above the level in question. In Fig. 4–8(a) we have represented two sea-level locations by 1 and 2. The masses of air above unit

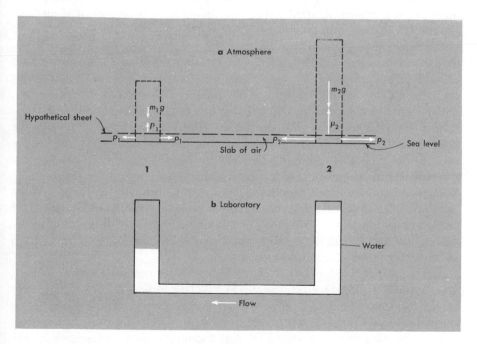

FIGURE 4-8   *Pressure forces in the atmosphere and in the laboratory.*

Winds of global scale

areas at each location are m_1 and m_2. The downward forces due to gravity are therefore $m_1 \times g$ and $m_2 \times g$, where g is the acceleration due to gravity.

Now consider an imaginary surface parallel to the ground at sea level; the distance above the ground is so small that the mass of air below the surface is negligible. The pressure at this surface acting upward on the two columns of air must be $p_1 = m_1 \times g$ and $p_2 = m_2 \times g$ if the air is not to experience a net force, and therefore to move either upward or downward.

It is, however, a property of a fluid that pressure is exerted equally in every direction. The inflation of a balloon is an example. If the membrane has equal strength at all points, the internal pressure produces a sphere. Thus the pressures p_1 and p_2 must also act sideways, as shown in Fig. 4–8(a). Consider the slab of air between the ground and our imaginary surface. Since p_2 is greater than p_1, this slab has a greater force acting from right to left than from left to right. The slab experiences a net force proportional to $(p_2 - p_1)$ that drives air from point 2 to point 1.

In order to assure ourselves that the horizontal pressure force does act this way, consider the laboratory analogy, a U-tube filled with water [see Fig. 4–8(b)]. The arguments used above can be applied to this situation also, and they lead us to expect a flow from right to left, as indicated. If there are any doubts about whether this expectation is correct, it is an easy experiment to carry out.

We may judge the acceleration that this pressure force will cause if the fluid starts from rest and no other forces act. According to Newton's laws of motion, force is equal to mass times acceleration ($F = ma$) as regards each individual element of fluid. The force is proportional to $(p_2 - p_1)$, but the mass of fluid undergoing acceleration is proportional to the distance from point 1 to point 2. Thus the acceleration will be proportional to the pressure difference divided by the distance. The term *pressure gradient* is used to describe this quotient. The pressure gradient can be readily pictured in terms of the degree of crowding of isobars on a weather map. Where isobars are close together, the pressure gradient is large; where isobars are far apart the pressure gradient is small. Air is accelerated from regions of high pressure to regions of low pressure, provided no other forces act on the air. We shall see, however, that there are many situations in which other forces must be considered.

Finally, we should note that the mass of air in our accelerating slab is proportional to the density. The ideal gas law indicates that the density of air is proportional to the pressure. Consequently, the acceleration is proportional to the pressure gradient divided by the pressure, rather than simply the pressure gradient. We are, therefore, more interested in the proportional difference in pressure between point 1 and point 2 than in the absolute difference. Suppose, for example, that the average pressure is 1000 millibars, and the difference in pressure between the two points is 10 millibars. Then the proportional pressure difference, equal to the absolute difference divided by the average pressure, is 1%. If, on the other hand, the pressure difference were 10 millibars, but the average pressure were only 500 millibars, the proportional pressure difference would be twice as large. The resulting acceleration of the air would, therefore, be twice as large in the second case as in the first case.

A Direct Thermal Circulation

Let us now consider the action of pressure forces on an atmosphere that is hotter in one region than it is in another. What we mean by a region should become clear as we consider two possibilities.

Venus rotates so slowly that one day on Venus lasts for 117 Earth days. We might reasonably expect that the nighttime atmosphere would become very cool, while the daytime atmosphere would be hot. In fact this does not happen to any great extent, and we shall see why, but the tendency obviously does exist and we may consider its possible consequences.

The second situation is that of the Asiatic monsoon. In the tropics during the summer the land becomes hotter than the neighboring ocean because of the greater capacity of the ocean to absorb and redistribute heat (a phenomenon that we considered in Chapter 3 when we were discussing surface temperatures on Mars).

Consider Fig. 4–9. We may appeal to observation to show that sea level pressure does not vary greatly. The map in Fig. 4–1 illustrates this. A few tens of millibars in a thousand are small compared to the proportional pressure difference that exists at high levels between the hot and cold regions. The reason is shown schematically in Fig. 4–9. According to the barometric law, the logarithm of the pressure decreases with height at a rate inversely proportional to the temperature. Thus, at the level marked A in Fig. 4–9, the pressure over the hot land is greater than the pressure over the cold sea, even though the ground level pressures are almost the same. In fact, the proportional change of pressure at a fixed height, $(p_{hot} - p_{cold})/p_{average}$, increases indefinitely the higher we go. According to our discussion in the previous section, the atmosphere will experience an increasing acceleration from hot to cold, or from land to sea, as altitude increases. The heavy arrow shows the resulting motion.

We must now consider continuity of the fluid. For our present purposes we may regard air as indestructible. Therefore, a flow into any region must be compensated by an equal flow out of the region. If this were not so, the amount of air in the region would increase indefinitely. At the point indicated by X in Fig. 4–9, continuity can be preserved simply by allowing the wind to flow straight on, in one side and out the other side of each

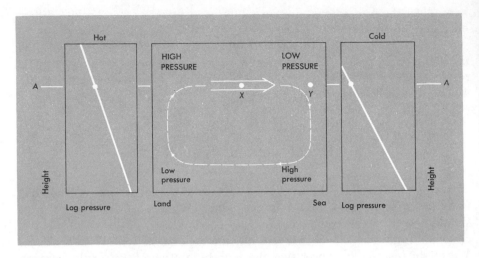

FIGURE 4-9 *A thermal circulation. The side panels show how pressure varies with height in the hot atmosphere (left) and in the cold atmosphere (right). The center panel shows the wind blowing from the high-pressure region (left) to the low-pressure region (right). The proportional pressure change is greater at the higher levels, and this is where the wind is strongest. There is a weaker return flow at the lower levels.*

small volume. At point Y, however, the situation differs. We have shown a rigid wall through which air cannot pass. The outflow from Y must, therefore, be either upward or downward. Here we are concerned with the downward flow, as shown in the figure.

As the air flows up to Y and turns the corner, it will clearly have an initial tendency to pile up and force more air to lie over the sea than over the land. At ground level the pressure will, therefore, be greater over the sea than over the land, and the pressure gradient will drive a wind flowing from sea to land. Fig. 4–9 shows that we could have predicted this flow from a continuity argument, as we predicted the downward flow at point Y. We see also that the complete circulation has hot air rising over the land and cold air sinking over the ocean. Thus the flow shown in Fig. 4–9 satisfies all of our intuitive ideas. Our reasoning also suggests that, under these conditions, there will be a low-pressure area over the land. This phenomenon is called a *heat low*; it is exhibited most strikingly by a huge low-pressure area that develops over Siberia during summer.

At this point the reader may wonder why we invoked a rigid vertical boundary to explain the downturn of the flow at Y, since the atmosphere clearly has no vertical boundaries. What the atmosphere does have is another hot region somewhere to the right of the figure, and a flow that is the mirror image of the one we have shown, repeated to the right. Two

Winds of global scale

streams of air will, therefore, converge at Y, and in practice the flow will be the same as if there were a rigid boundary.

The circulation we have described is often called a *Hadley cell*, named after the person who first proposed the existence of this kind of circulation in the Earth's atmosphere. It is possible to make a laboratory model of the Hadley cell by using a pan of water heated from below at one end and cooled from below at the other end. Aluminum powder suspended in the water makes the flow visible because the little particles of aluminum are carried along by the water. Figure 4–10 is a photograph of this experiment. A fairly long exposure has been used so that the moving particles appear as streaks. The streaks are long where the particles are moving fast, and they are short where there is little movement. The photograph shows that the flow is far from being symmetrical about the center of the pan. Instead, there is a narrow column of rapidly rising water concentrated at the hot end, but over the rest of the pan the water is descending slowly. This asymmetry of the Hadley circulation may have important consequences on Venus.

Circulation in the Atmosphere of Venus

We know very little about the planet Venus. The main reason for this is that the planet is covered by clouds that are so thick that we cannot see down to the surface with any kind of radiation other than centimeter radio waves. The cloud top lies at a height above the solid surface of about 65 km, where the pressure is about 100 millibars and the temperature is close to 250°K. At the surface of Venus the mean temperature is 700°K, and the pressure is 80 atmospheres (approximately 80,000 mb). We have no direct information of any kind about motions below the cloud tops, but we can see changes taking place in the appearance of the clouds, which can only be associated with winds.

Figure 4–11 schematically shows one theory of the deep circulation on Venus. Planetary emission follows the Stefan-Boltzmann law (Flux $= \sigma T^4$) and is observed to vary little over the surface of the planet. The input of solar radiation varies from large to small values; there is a net gain of heat in some places and a net loss of heat from the planet to space in others. In this model it is supposed that all of this radiative heat exchange takes place solely at the cloud tops and that no solar radiation penetrates into the clouds. This may not be true, but some of the deductions from the model do not depend upon this assumption.

The circulation in Fig. 4–11 bears an obvious resemblance to that in Fig. 4–10. Its sense of rotation is reversed, but that is because heating and cooling occur along the top surface rather than the bottom. In both of these

Winds of global scale

HOT

COLD

FIGURE 4-10 Hadley cell circulation in water heated at the bottom right-hand corner and cooled at the bottom left-hand corner. Suspended aluminum powder makes the flow visible. (From H. T. Rossby, 1965, Deep-Sea Research.)

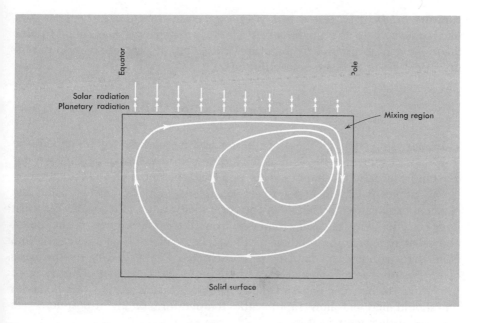

FIGURE 4-11 *Schematic representation of the possible circulation in the atmosphere of Venus. (After R. M. Goody and A. R. Robinson, 1966.)*

cases, and in the case of Fig. 4–9, the air rises where it is hot and sinks where it is cool, as, in a general way, we would anticipate.

Figure 4–11 is a schematic representation of an elaborate theoretical analysis. In the course of this analysis calculations were made of the effect of the motion upon the vertical and horizontal temperature structure. With regard to the latter it appears that the motions will tend to eliminate temperature differences caused by differences in insolation in accord with observations.

The theory suggests that the vertical temperature profile will resemble an adiabatic profile, with temperature increasing steadily as the altitude decreases. Descending air is compressed as it moves to lower levels in the atmosphere. The compression causes the temperature to increase, as we explained in Chapter 3. If the circulation is sufficiently rapid, and if the air does not cool too fast by emission of radiation, the temperature will increase at the adiabatic rate. This is precisely what is observed on Venus. Venera and Mariner Venus spacecraft have all found that the temperature increases adiabatically as altitude decreases in the lower atmosphere. As we explained in Chapter 3, this observation could also be the result of thermal convection driven by solar radiation deposited at the ground, but we cannot be sure that the radiation actually reaches the ground. What we are now suggesting as an alternative explanation is that the adiabatic temperature gradient is related to a planetary circulation driven by heat supplied unevenly to the upper levels of the atmosphere. According to this theory, the high ground temperature is caused, at least in part, by compressional heating of the descending air.

The Equatorial Hadley Cell

In our discussion of the Earth's general circulation we identified the tropical cell shown in Fig. 4–6 as a Hadley cell, thus implying that we understand the basic mechanism. This point can be debated, for the situation in the tropics is extremely complicated. Let us suppose, however, that the classical explanation of Hadley is correct, at least in some respects. We must now consider why the cell circulates from the equator toward the poles, and what influence the rotation of the planet will have on such a motion. Note that, without explanation, we constructed Fig. 4–11 also on the basis of an equator-to-pole circulation, although the most obvious differences of insolation are between day and night. In order to understand this behavior we should find out how rapidly the temperature of an atmosphere can respond to changes in heat input.

Using results from Chapter 3, let us calculate how much the atmospheric temperature would change during a day if all available solar energy were deposited in the atmosphere and no energy at all were lost. The total heat input is given by Eq. (3–1),

$$Q = \pi R^2 S \, (1\text{-}A) \tag{4–1}$$

where R is the radius of the planet, S the flux of solar radiation, and A the albedo. We get the total amount of heat absorbed during a day by multiplying Q by the length of the day, t_d. To determine how great a temperature change this will cause, we first calculate the mass of the atmosphere per unit area of surface; this is p_0/g, where p_0 is the pressure at the ground and g is the acceleration due to gravity. The heat required per unit area of surface to change the temperature of the atmosphere by 1°K (the *heat capacity*) is therefore, $p_0 c_p/g$, where c_p is the specific heat at constant pressure. To find the heat capacity of the entire atmosphere we multiply this quantity by the surface area, $4\pi R^2$. The heat required to raise the temperature of the atmosphere by an amount ΔT is given by the product of the heat capacity and the temperature increase, $4\pi R^2 p_0 c_p \Delta T/g$. Now we may estimate the temperature increase during a day by equating the heat required to the heat available

$$4\pi R^2 p_0 c_p \Delta T/g = Q t_d = \pi R^2 S (1\text{-}A) t_d \tag{4–2}$$

Rearranging and dividing both sides by the effective temperature, T_e (see Table 3–1), we find

$$\frac{\Delta T}{T_e} = \frac{S(1\text{-}A)gt_d}{4p_0 c_p T_e} \tag{4-3}$$

Most of the data needed to calculate this ratio for the different planets have been presented in Tables 1–5, 3–1, and 3–2. For Earth, where the surface pressure is 1013 mb (1 mb = 1000 dyne cm^{-2}) and the length of the day is 8.64×10^4 sec, we find that $\Delta T/T_e$ is 0.68%. This means that there will be little difference in temperature between the day and the night side; the atmosphere has sufficient heat capacity to absorb all of the solar heat incident during the course of a day with little change in temperature, in effect averaging the heat input over both day and night.

At first this statement might appear to be incorrect. Experience tells us that it is hotter near the ground during the day than during the night, particularly in dry areas such as deserts. These changes, however, involve only a small proportion of the whole atmosphere and have little global influence. A weather forecaster discussing weather maps of the continental United States need not distinguish between day and night. He does not, for example, expect to find large wind systems shifting direction from easterly by day to westerly by night. Such changes may happen in certain limited areas and over a limited height range (coastal sea breezes are an example) but they are not a prominent feature of the atmosphere as a whole.

On Jupiter we expect even less diurnal variation than on Earth, largely because the planet is so far from the Sun that the flux of solar heat [S in Eq. (4–3)] is very small. Taking p_0 to be greater than the cloud top pressure of 1000 mb, we find that $\Delta T/T_e$ is less than 1.5×10^{-3}%. On Mars the surface pressure is only 6 mb, and we calculate a value of 38% for $\Delta T/T_e$. We can, therefore, predict significant diurnal variations of temperature and wind in the atmosphere of Mars.

For Venus we calculate $\Delta T/T_e = 0.36$%, little different from the value for Earth. The high surface pressure on Venus (80 atmospheres) tends to cancel the effect of the long Venus day (117 Earth days). Thus, we do not anticipate large day to night temperature differences on Venus and we therefore do not anticipate a Hadley cell rising on the day side and sinking on the night side. The important difference in solar radiation is between equator and pole, as was shown in Fig. 4–11.

The variation with latitude of the input of solar radiation requires further examination. Figure 4–12 illustrates the situation for Mars and Earth, both of which have rotation axes about equally tilted with respect to the planet-sun direction. We have shown that the solar radiation behaves as if averaged over both day and night. At a given latitude the average insolation depends on two factors—the relative length of day and night and the flux of solar radiation per unit area of surface.

The figure shows how beams of sunlight with equal energy fluxes dis-

Winds of global scale

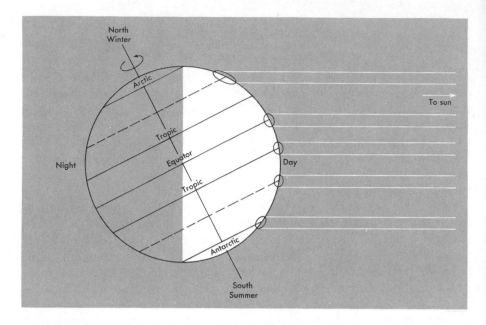

FIGURE 4-12 *Incident solar radiation at a solstice. A beam of sunlight is spread over a larger area of ground at high latitudes, where the Sun is close to the horizon, than at low latitudes where the Sun is almost overhead. The day is longer than the night in the summer hemisphere whereas the night is longer than the day in the winter hemisphere. Both effects are important in determining the incident solar radiation.*

tribute their energy over different areas of surface at different latitudes. The energy is spread over the smallest surface area at the latitude where the Sun is overhead at midday (the southern tropic in Fig. 4–12). At noon, the solar energy falling on unit area of surface is, therefore, greater at this latitude than anywhere else on the globe.

If the planet had no axial tilt, as is approximately the case for Venus, day and night would be equal at all latitudes, and only at the equator would the Sun ever be directly overhead. The average solar flux would, therefore, be a maximum at the equator and a minimum at the two poles. Under this circumstance it is a simple matter to show that the average solar flux is proportional to the cosine of the latitude; the cosine of the latitude is unity at the equator and zero at the two poles.

The axial tilt complicates this problem because the day is now longer in the summer hemisphere than in the winter hemisphere (compare the relative shaded and unshaded portions on a single latitude circle in Fig. 4–12). Allowance for both the variation with latitude in the length of the day and the variation in the angle at which sunlight is incident on the ground leads to results shown in Fig. 4–13. The lines in this figure join points at which the solar flux, summed over a day, is the same. The most surprising feature of the results is that the maximum occurs not in the equatorial regions but in the polar summer because there the day lasts for twenty four hours. On this basis alone we might expect the highest atmos-

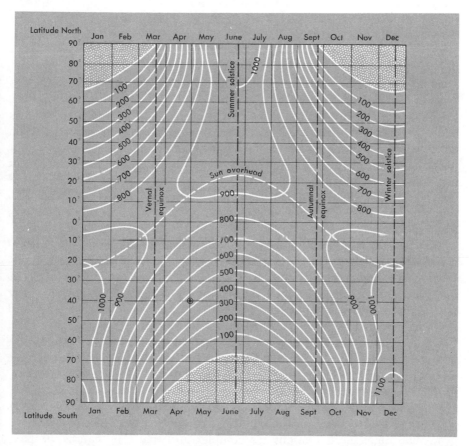

FIGURE 4-13 *Incident solar radiation for Earth (contours of cal cm^{-2} day^{-1}) as a function of latitude and date. Stippled areas represent latitudes with continuous night. The figure shows that there is more radiation incident on the summer pole than on the tropics. However, if allowance is made for absorption and scattering of the radiation as it passes through the atmosphere, the situation is reversed and, at the ground, the tropics receive slightly more radiation than the summer pole. (In order to understand the use of this diagram consider May 1st at latitude 40°S. The operating point is marked ◎ . From the contours of constant insolation we find that 450 calories fall on 1 cm^2 each day. One calorie equals 4.18 × 10^7 ergs.) (From A. Miller, 1971, Meteorology, Charles E. Merrill Publishing Co.)*

Winds of global scale

pheric temperatures to occur during polar summer. If this were the case, the rising branch of the Hadley cell would be over the pole and not over the equator. This conclusion is not in accord with observation on Earth, but it may be correct for Mars, which has the same axial tilt as Earth; this point will be explored more fully later in the chapter.

For Earth, however, we must not forget the albedo, since much of the incident solar radiation is reflected off into space rather than absorbed (and only absorbed radiation influences the temperature). On Earth the dense atmosphere ensures a large amount of scattering from the slanting rays of sunlight in the polar regions. Snow and ice, and general cloudiness also, tend to increase the albedo near the poles, so that the product, incident solar flux times one minus albedo, is greater in tropical regions both in summer and in winter. According to Eq. (3–4) the tendency is, therefore, for the highest temperatures to occur in the tropical regions, as is observed. It is nonetheless true that the temperature difference between equator and pole is much less in the summer hemisphere than in the winter hemisphere. If we relate temperature differences to pressure differences and consider the discussion of thermal circulations and pressure forces that appeared earlier in this chapter, we may anticipate greater pressure differences and, therefore, greater accelerations of the air in winter than in summer.

In a general way, we now have the information to account for the tropical Hadley circulation on Earth. The large heat capacity obscures the distinction between day and night. The solar heating averaged over a day is largest in the tropics, smaller at the summer pole, and least at the winter pole. By analogy with the laboratory experiment shown in Fig. 4–10, and recalling our own theoretical discussion, we anticipate a rising current in the tropics and a sinking current at higher latitudes, with an equator-to-pole motion high in the atmosphere and a pole-to-equator motion near the ground. We anticipate that this cellular circulation will be stronger in winter than in summer.

Our anticipations correctly describe the Earth's tropical Hadley cell, as shown in Fig. 4–6 between the equator and 30°N. Let us not forget, however, that atmospheric motions are complex and that the analysis we have given of the forces and processes responsible for the tropical circulation is not the only possible one. The ideas we have discussed may be no more than a framework containing some elements of the truth.

The Coriolis Force

We have established that the Earth's global Hadley circulation will be in a north-south direction, perpendicular to the surface velocity of the rotating planet. Let us now consider a small parcel of air with a mass of 1 gram. Initially the parcel is located at the equator (Fig. 4–14, points A and A'). If the planet has a radius R cm and an angular velocity of rotation Ω radians per second, the angular momentum of this parcel is, as a matter of definition, $\Omega \times R^2$. The sense of the angular momentum is that of the Earth's rotation, from west to east, which is called westerly.

Suppose that the parcel is moved from A to B (see Fig. 4–14) by a Hadley circulation, without experiencing any forces in the east-west direction. It is a consequence of Newton's laws of motion that the angular momentum of the parcel will remain constant. The parcel, however, is held close to the planet's surface by the force of gravity, and its distance from the axis of rotation has changed from R to $R \cos \phi$ where ϕ is the latitude (see Fig. 4–14); thus, the angular velocity of the air must change if the product of (radius)^2 and angular velocity is to remain constant. If the new angular velocity of the air is Ω' we can write

$$\Omega' (R \cos \phi)^2 = \Omega R^2 \tag{4–4}$$

This is the statement of the law of conservation of angular momentum. Since $\cos \phi$ is at most equal to unity, Ω' is greater than Ω, the surface angular velocity. To an observer at B', therefore, the parcel of air from A, A' will appear to rotate more rapidly than the planet and to have a westerly wind velocity with respect to the surface.

We can reverse the argument and consider another parcel starting at B with the angular velocity Ω of the surface at B'. When it reaches A, it will have an angular velocity Ω', less than that of the surface at A'. Thus, it will

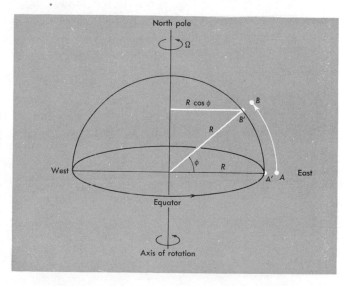

FIGURE 4-14 *Geometry of the Northern Hemisphere and the conservation of angular momentum. The angular momentum of unit mass fixed to the earth's surface is less at B' than it is at A'. Away from the surface unit mass moving from A to B tends to maintain its angular momentum.*

Winds of global scale

appear as an easterly wind with respect to the surface. We begin to see why the wind systems on Earth have strong east-west components; at the same time it is becoming clear that the motion of air on a rotating planet may be complicated.

An interesting aspect of the laws of physics is the manner in which they may depend on the observer's *frame of reference*. Newton's laws of motion require a particle to continue in a state of rest or of uniform motion in a straight line unless acted upon by an external force. This statement, however, depends on the frame of reference. It refers to an *inertial frame*, one in which the universe as a whole appears to be at rest. This does not correspond to the surface of a rotating planet. Viewed from the Earth's surface the fixed stars appear to rotate around the north pole and to move horizontally from east to west.

The horizontal motion does not concern us, for Newton's laws do not distinguish between a state of rest and a state of uniform motion in a straight line. An observer inside a smoothly riding railway coach, travelling at constant velocity, would not be able to find anything wrong with our statement of Newton's law if he made experiments in this moving laboratory. His experience would be completely different if the coach were to rotate.

Figure 4–15 represents a view looking down on the north pole. Earth rotates anticlockwise with angular velocity Ω. For the sake of simplicity we may picture it as a flat disk rather than a sphere. This simplification will not affect the essence of our discussion. A projectile is now thrown from the north pole in the direction OD. In Fig. 4–15(a) we see the motion in an inertial frame, that of an observer above the rotating disk whose position is fixed, we shall suppose, with respect to the universe. The projectile travels with constant velocity in a straight line, and in equal time intervals, say one second intervals, it will successively reach the points A, B, C, and D. While the projectile is travelling, however, the disk rotates, and the points which were initially at A, B, C, and D move to new positions A', B', C', and D'.

Now consider a frame of reference fixed on the disk. Suppose the observer cannot see anything but the disk.* The points O, A', B', C', and D' appear to him to remain in a straight line. Consequently, he concludes [Fig. 4–15(b)] that the projectile follows a curved path.

The observer in the rotating frame now has a problem with Newton's laws: the projectile is not acted upon by a force while travelling through the air, and it should, therefore, move uniformly in a straight line. He must choose between abandoning Newton's laws and modifying them. In prac-

*The German fluid dynamicist, Ludwig Prandtl, built in his laboratory a closed sentry box on a large rotating wheel. A student was secured inside and given a rubber ball to play with (he had to be secured in order to be safe). At the conclusion of this section the reader may wish to imagine what might happen as the ball is bounced on the floor or thrown across the box.

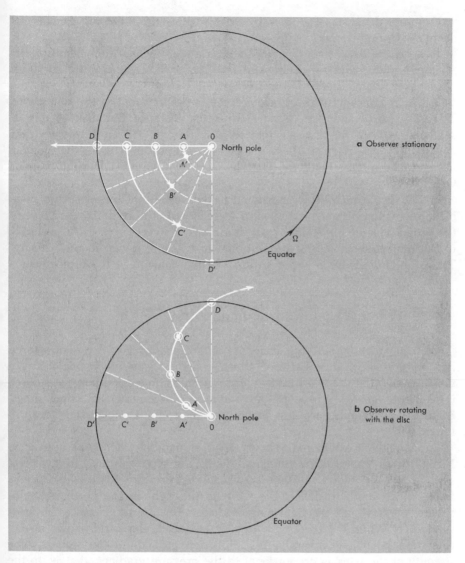

FIGURE 4-15 *The Coriolis force. A projectile is thrown in the direction OD in (a). Its subsequent motion appears to depend upon whether the observer is stationary or rotating with the disk.*

tice it is more convenient to modify the laws by assuming the existence of a force that generates the curved trajectory. This new force is called the *Coriolis force*. If it is included in Newton's laws, it enables us to use these laws without further reference to the fact that we live on a rotating planet.

Winds of global scale

From our discussion of Fig. 4–15 we may draw the following conclusions about the Coriolis force:

1. It acts to deflect a projectile to the right in the northern hemisphere. A simple extension of our discussion would show the reverse in the southern hemisphere, that is, deflection to the left.

2. The magnitude of the Coriolis force is proportional to the deflections DD', CC', BB', and AA'. Now DD' is the distance the disk rotates in 4 seconds, CC' is the distance rotated in 3 seconds, BB' in 2 seconds, and AA' in 1 second. Each of these distances is proportional to Ω, the angular velocity of the disk. Thus, the Coriolis force is proportional to the angular velocity.

3. The distance OD is proportional to the velocity of the projectile since it represents the distance travelled in 4 seconds. For a fixed angular velocity the proportions of Fig 4–15(a) are not affected by changes in the velocity of the projectile, so DD' is proportional to OD and hence to the velocity. Thus, the Coriolis force is proportional to the velocity of the projectile.

A Balanced Wind

We can now combine our concepts of Coriolis force and pressure force and consider a balance between them, called a *geostrophic balance.* We must assume that the wind is steady, not accelerating, for otherwise a force is needed to create the acceleration. We must also assume that other forces, friction for example, are less important than the Coriolis and pressure forces.

Suppose the weather map shows a set of parallel isobars, as appear in Fig. 4–16(a). The pressure force is indicated; it acts from high to low pressure. If a wind blows at right angles to the pressure force, the Coriolis force, which tends to deflect the wind to the right in the northern hemisphere, can have just the right magnitude and direction to balance the pressure force. When the forces are exactly balanced, the wind shows no acceleration but blows steadily in a direction parallel to the isobars. The strength of the wind is proportional to the pressure gradient, that is, to the crowding of the isobars.

As regards the global circulation, we have argued that the tropics, where more solar energy is absorbed, are hotter than are high latitudes. Following the argument of Fig. 4–9 then, the pressure force at high levels is directed from the equator towards the pole (northward in the northern hemisphere). On examining Fig. 4–16(a) we see that the corresponding balanced wind is westerly.

This wind direction agrees with the result of the argument we gave earlier, based on conservation of angular momentum, when we started with a

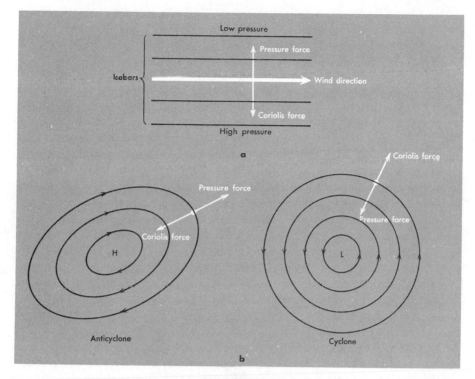

FIGURE 4-16   *Geostrophic balance in the Northern Hemisphere. The lines represent isobars (lines of constant pressure). The balanced wind blows along the isobars.*

package of air in the tropics and transported it northward. It also agrees with the wind directions at high levels indicated in Fig. 4–5. In fact, the wind at high levels in middle latitudes in the Earth's atmosphere is very nearly balanced in the way we have described. The same result is true for the southern hemisphere. The Coriolis force is reversed because the direction of rotation looking down on the pole is the opposite of that in Fig. 4–15. But the pressure force in the southern hemisphere is directed from north to south, which brings us back to a westerly balanced wind.

Finally, and somewhat beside the main point of this chapter, we may remark that winds can be in geostrophic balance in weather systems such as the cyclone and anticyclone shown in Fig. 4–16(b). This pattern of winds, clockwise around an anticyclone and anticlockwise around a cyclone, is characteristic of most weather maps. This behavior was noticed before it was understood and is incorporated in *Buys Ballot's law*, which asserts that the higher pressure will be to the right, facing with the wind, in the northern hemisphere. The reverse is true in the southern hemisphere.

Winds of global scale

Circulation of the Martian Atmosphere

Our ideas on thermal circulations and Coriolis forces seem to be consistent with observation in the Earth's atmosphere, but if they are really sound, they should also account for motions on other planets, for example, on Mars. Unfortunately, we do not have detailed wind measurements on Mars, but we do have numerical models based on the methods used to predict weather on Earth with the aid of electronic computers.

It may seem unfair to compare two theories, one simple and one extremely complex, as a test of their accuracy. This is not, however, as unreasonable as it seems, for the numerical methods lead to such complicated results that it is almost as difficult to understand a model of the atmosphere as it is to understand the atmosphere itself. In either case, the description of phenomena in terms of simple concepts is the object of our discussion. Moreover, terrestrial experience has shown that the average behavior of a numerical model is remarkably close to the average behavior of the actual atmosphere, provided the essential properties of the atmosphere are included in the calculation.

The calculation that we shall describe is made by standard methods of numerical weather prediction, with the parameters and some of the physical mechanisms changed from those appropriate for Earth to those appropriate for Mars. In numerical weather prediction we suppose that the properties of the atmosphere (wind velocities, temperature, and pressure) are known at a large number of points at which we perform the calculations. The points may be spaced, for example, by 500 km in both east-west and north-south directions. In addition, we approximate the vertical structure of the atmosphere by a series of discrete layers.

The complete set of mathematical equations for all of the physical processes involved are written in a form that relates the physical variables at one point to those at neighboring points. From these equations it is possible to calculate, by arithmetic operations, the rate at which the physical variables change. Thus, if we know their values at one time, we can calculate their values a short time later (a *time step* later) and thereby determine the future state of the atmosphere. By repeatedly stepping forward in time, we can predict the physical state of the atmosphere 1 day or 2 days ahead, or, if we choose, 100 or 200 days ahead. The value of the prediction is usually measured by the accuracy of the details, particularly those related to weather elements. This accuracy decreases with time from exact knowledge of the initial state to no better than a general description of the atmosphere after many days. There are hopes that detailed predictions will eventually be made as far in advance as 10 days but, for the present, weather services are content to use 24 and 48 hour forecasts made by these methods.

When we apply the methods of numerical weather prediction to Mars, we are no longer interested in the details of weather systems. The calculation is started from a reasonable guess of the state of the atmosphere and carried forward in time until all relationship to the initial state is lost. For Mars this takes between 10 and 20 days. The expectation is that, after this time has elapsed, we will obtain a plausible, albeit fictitious, weather map for the planet. Results of a calculation of winds and average temperatures for this make-believe world at a single instant of time during the southern summer are shown in Fig. 4–17.

Before discussing these results, let us summarize our conclusions from previous sections. Figure 4–13 suggests that for the relatively transparent Martian atmosphere (Mars has a mean pressure of only 6 millibars and very few clouds) the highest insolation and highest temperatures may occur in the polar regions at the summer solstice; the lowest temperatures may be expected in the polar winter. Figure 4–17 (right panel) shows just these features. From the discussion of Fig. 4–9, we expect pressure forces aloft from equator to pole in the winter hemisphere and from pole to equator in the summer. Finally, since Mars rotates rapidly, we may expect strong westerlies in the winter hemisphere and weak easterlies in the summer hemisphere, assuming that the wind is approximately balanced (Fig. 4–16).

These predictions, both with respect to the temperature and the wind, are realized in Fig. 4–17. The winds in the northern hemisphere are irregular but, as we find with terrestrial mid-latitude winds, they are generally westerly. In the southern hemisphere weak easterlies prevail.

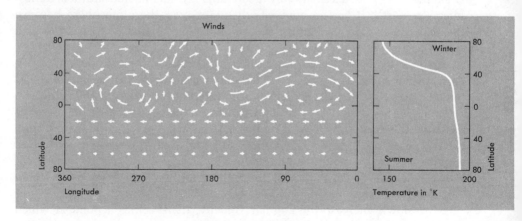

FIGURE 4-17 *A theoretical weather map for Mars. Calculations refer to the atmosphere well away from the ground. On the right is the variation of temperature with latitude. On the left is a map of the wind on a given day during southern summer. (After C. Leovy and Y. Mintz, 1969.)*

Winds of global scale

Planetary Waves

The northern hemisphere winds in Fig. 4–17 suggest a series of waves travelling from west to east and encircling the planet. Planetary scale waves are important on Earth not only because they relate to weather systems (the cyclones and anticyclones and associated structure of importance to the weather forecaster) but also because they can be regarded as a kind of turbulence that must be understood if we are to describe completely the average motions of the atmosphere. The terrestrial weather map shown in Fig. 4–18 shows four huge waves meandering around the pole. This map was selected because the wave effect is particularly obvious; waves are not so easily seen in Fig. 4–1. Figure 4–1 refers to sea level: the flows and pressure patterns are characteristic of the lower part of the atmosphere; Fig. 4–18, on the other hand, refers to a higher level of the atmosphere. Figure 4–19 shows a theoretical and highly schematic interpretation of how a wave pattern at high levels can be associated with cyclones and anticyclones in the lower atmosphere.

Much attention has been given to the explanation of these wave-like patterns in the atmosphere. They can be reproduced with particular clarity in laboratory experiments. Figure 4–20 illustrates the apparatus used to obtain the pictures reproduced in Fig. 4–21. A fluid, usually water, is contained between two cylinders that are maintained at different temperatures. The whole assembly is rotated at constant angular velocity. This experiment is simple, but it is a remarkably good analogy to the mid-latitude atmosphere on Earth. It has an obvious advantage over the atmosphere because the experiment can be controlled and the consequences of changes in a single parameter can be studied.

Suppose we fix the temperature difference between the cylinders while we increase the speed of rotation. When the angular velocity is small, we find that the water rotates at a speed slightly different from that of the turntable. In other words, there is a wind in the frame of reference of an observer fixed on the turntable. The direction of the wind is as anticipated from the foregoing discussions. The situation in Fig. 4–20 resembles the Martian summer with a hot pole; we anticipate an easterly wind, that is, a flow rotating more slowly than the walls of the annulus. This flow is shown in Fig. 4–21(a).

If we slowly increase the angular velocity, there comes a point at which the flow changes from this regular symmetric pattern to a pattern showing waves [see Fig. 4–21(b)]. The number of waves can be relatively large if the temperature difference is small; five or six have been recorded. The important point is that a critical condition is reached at which the flow

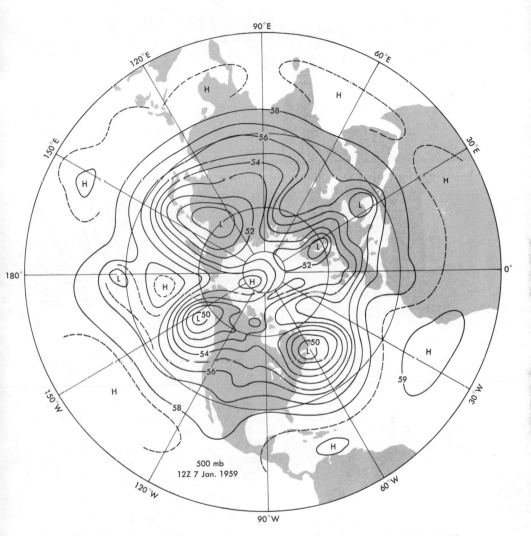

FIGURE 4-18 *Northern Hemisphere 500 mb chart for 7 January, 1959, 1200 GMT. Heights are expressed in units of 100 m. See Fig. 4-4 for an explanation of the contours. (From E. Palmén and C. W. Newton, 1969, Atmospheric Circulation, Academic Press.)*

changes its nature. This is analogous to the sudden onset of turbulence in Reynolds' experiment, shown in Fig. 4–7.

It is clear that once waves develop, they will have important effects on the mean flow, as was true for turbulence in Reynolds' experiment. The waves represent a flow moving from the inner to the outer cylinder and back, at relatively high velocities. The fluid is heated when close to the inner cylinder and cooled when close to the outer cylinder. This flow should

Winds of global scale

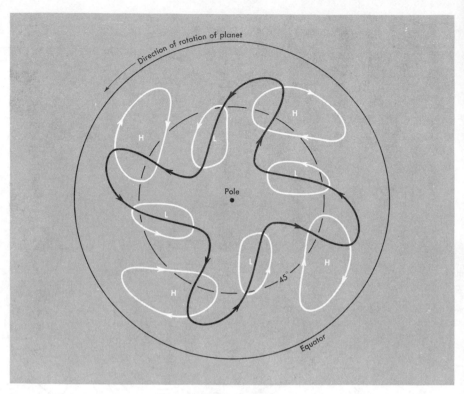

FIGURE 4-19   *Schematic representation of planetary waves. The heavy line represents the wave-like flow in the upper levels; the lighter lines, forming closed patterns are the associated cyclones and anticyclones in the lower atmosphere. From Y. Mintz, 1961, in The Atmospheres of Mars and Venus, Publication 944, Ad Hoc Panel on Planetary Atmospheres of the Space Science Board, National Academy of Sciences-National Research Council, Washington, D. C., 1961.)*

be efficient at transferring heat from the inner to the outer cylinder. If only a limited amount of heat were available, it could have the effect of decreasing the temperature difference between the cylinders, possibly to the point where the symmetric, smooth flow would once again become the preferred motion. This illustrates how wave motions, and also the less regular motions that occur in the lower atmosphere, can have profound effects upon the mean flow.

Retrospect

The final conclusion of the last section is not new; it was reached at an early stage of our discussion when we wished to examine the mean atmos-

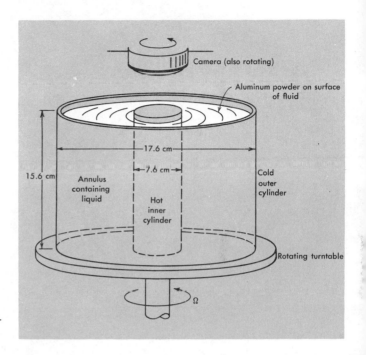

Camera (also rotating)

Aluminum powder on surface of fluid

17.6 cm

7.6 cm

15.6 cm

Annulus containing liquid

Cold outer cylinder

Hot inner cylinder

Rotating turntable

Ω

FIGURE 4-20 *Hide's annulus experiment.*

pheric motions, ignoring the weather and other irregular phenomena. We started by making a case for extracting the mean motions and neglecting weather, and we have come to the point of predicting the existence of the irregular motions that we tried to average out. The chain of ideas was: to establish the existence of pressure forces; to show that these must exist where temperature differences exist; to find that they lead to a direct thermal circulation or Hadley cell; to examine the effect of rotation upon such a circulation (pausing by the way to consider Venus), and thus to demonstrate the existence of strong east-west flows; to discover, by laboratory analogy, that such flows are unstable if the rate of rotation or the temperature difference between equator and pole is sufficiently large; and hence to return to the weather systems. The chain is self-consistent and complete in essentials.

The complicated processes of weather can be regarded as an outgrowth of the regular planetary waves we have described. This is illustrated schematically in Fig. 4–19. The flow in the upper levels looks like the laboratory flow in the annulus; at lower levels, however, the flow breaks into closed cells, still associated with the wave-like flow aloft, but identifiably cyclones and anticyclones.

We have tried to rationalize each step in our argument in order to discern the basic physical mechanisms behind motions in planetary atmospheres. Fortunately, we are not completely dependent on such arguments.

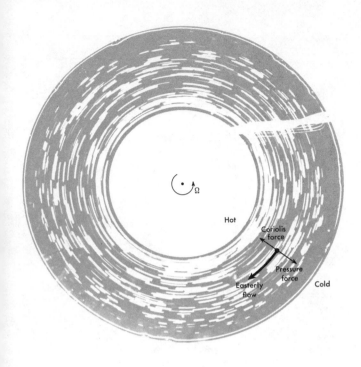

FIGURE 4-21 *Flow patterns in the annulus experiment. In both cases the inner cylinder is at 25.8°C and the outer is at 16.3°C. The heavy arrows indicate the motion of the fluid viewed in the rotating frame of reference. (Photographs by R. Hide, courtesy of the Royal Meteorological Society.)*

(a) Symmetrical zonal flow at a low rate of rotation, $\Omega = 0.341$ radians sec^{-1}.

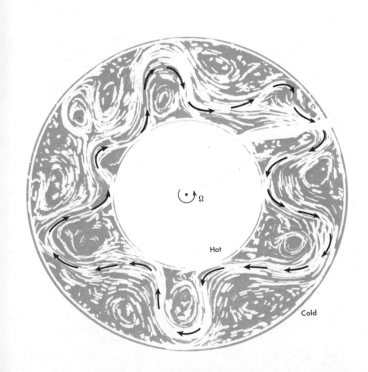

(b) Unsymmetrical and variable wave motions at a high rate of rotation, $\Omega = 5.02$ radians sec^{-1}

Winds of global scale

As we saw for Mars, numerical methods using electronic computers are available, and some might argue that the best road to understanding is through the use of numerical calculation. Undoubtedly, numerical simulation of important atmospheric processes is possible. It can happen, however, that the simulation is hardly less complicated than the atmosphere itself, and the chain of cause and effect can be almost impossible to follow. An intuitive understanding of the key processes is vital if we are to apply our ideas to other planets where detailed data are not available. Attempts to understand such processes in simple terms are as important in the computer age as before; it is just that our approach to these inherently complicated phenomena is likely to be influenced by the computational facilities now available.

Winds of global scale

5

Condensation and clouds

Clouds and the Radiation Balance

We have not yet considered the consequences of the fact that each molecular species can exist in different phases— as a gas, as a liquid, or as a solid. The physical properties of a given species vary greatly from phase to phase. Specifically, the effect of a species on solar and planetary radiation depends on whether it is gaseous, liquid, or solid.

Consider water on Earth. As a gas it is largely opaque to planetary radiation and relatively transparent to solar radiation (see Fig. 3–4), giving rise to an important greenhouse effect. As a cloud, the same amount of water is even more opaque to planetary radiation while, at the same time, it reflects most of the solar radiation back to space. If the ice or water is on the surface, there is again a difference between phases. Water absorbs about as much solar radiation as does damp earth, but snow or ice reflects strongly, thereby keeping the surface cool.

Although water on Earth is the prime example of a condensable substance that we shall discuss, clouds may be even more important on Venus, Jupiter, Saturn, and perhaps others of the outer planets as well. These planets are almost entirely covered by clouds of substances other than water and ice. Clouds are relatively rare on Mars, but the condensation of carbon dioxide at the poles may control the total amount of gas in the atmosphere.

One of the most significant effects of clouds is that they interfere with solar radiation and scatter more energy back into space than does clear air. In Chapter 3 we calculated the effective temperature of the Earth on the basis of an albedo (average reflectivity) of 0.33 and found a value of $T_e = 253°K$. Most of that reflectivity is caused by clouds, as is illustrated in Fig. 3–5, even though clouds, on average, cover only about 50% of the surface of the globe. Let us suppose, by way of illustration, that our planet is completely covered by clouds similar to those on Venus. The albedo in this case would be 0.71 (see Table 3–1), and the effective temperature would be 205°K. If, on the other hand, there were no clouds at all in our atmosphere and the albedo were that of Mars (0.17), the effective temperature would be 267°K.

Thus, a change from no clouds at all to complete cloud cover could decrease the effective temperature by 62°K. Under the assumption of a greenhouse mechanism, the same proportionate change in the fourth power of temperature would take place at every level in the atmosphere. Conditions on Earth would be drastically affected. If average ground-level temperatures were to rise by 16°K from the present day value of 288°K to 304°K, corresponding to a change in T_e from 253°K to 267°K, the water vapor pressure would increase about three times. Additional water vapor would enhance the greenhouse effect and further increase the ground temperature. The polar caps would melt, raising the level of the oceans by 15 m, flooding part of the continents, and drastically changing the climate.

The importance of clouds to all of the phenomena discussed in Chapter 3 and Chapter 4 cannot be overestimated, but their mode of action or coupling to these other phenomena is relatively complicated. Clouds are ephemeral. Left to themselves, cloud particles would fall to the ground and have relatively little effect on the atmosphere. Clouds, therefore, must be continually created, and this creation process usually involves motions of the atmosphere. Once formed, the clouds influence atmospheric temperatures, which in turn influence atmospheric motions. Even this brief examination reveals three processes that we have discussed in separate chapters, but which are really parts of a single, interlocking system.

Saturated Vapor Pressure

Consider the surface of a lake or a sheet of ice and the atmosphere above it. Water molecules continuously detach themselves from the water or ice surface and evaporate into the atmosphere. The rate of detachment

Condensation and clouds

depends upon the thermal agitation of the molecules and is determined by the temperature.

If evaporation were the only factor to be considered, the ice or water would in time disappear completely into the vapor phase. This process can take a long time if the rate of evaporation is slow enough. For example, it is believed that Saturn's rings consist of ice, but at a temperature so low that they could not evaporate in the lifetime of the solar system. For practical purposes, ice near Saturn is as permanent as is granite at typical terrestrial temperatures.

It is a matter of common experience, however, that water and ice evaporate readily at normal atmospheric temperatures. Thus, if oceans or bodies of ice are to persist for long periods of time, they must receive as many molecules by condensation as they lose by evaporation. The water or ice is then in a state of equilibrium with water vapor in the atmosphere.

The rate of condensation depends on the rate at which molecules impact the surface, and this rate depends on the partial pressure* of the water vapor. A state of equilibrium, therefore, prevails between ice or water and water vapor at a given temperature only if the vapor pressure has the right value. When this pressure of water vapor exists, we say that the air is *saturated*.

Figure 5–1 shows the relationship between saturated vapor pressure and temperature. We have used degrees Centigrade ($^\circ$C) instead of degrees Kelvin ($^\circ$K = $^\circ$C + 273.2) for this figure because of the importance of the melting of ice at 0°C. The division between ice and liquid water is indicated by the vertical line TZ. Saturated vapor pressures are given by the lines XTY' for water and TY for ice. Along these lines water vapor is in equilibrium with either ice or liquid. If conditions of temperature and pressure do not lie on one of the lines, either evaporation or condensation is more rapid, and the water molecules eventually exist in one phase only (we shall discuss below the broken line TY' referring to water below 0°C).

Suppose that the vapor pressure is 10 mb. If the temperature is +15°C, evaporation exceeds condensation and the water will all end up as vapor (point A). If the temperature is +3°C, condensation onto a water surface exceeds evaporation and the vapor will end up in liquid form (point B). At −10°C, the water all turns to ice (point C).

Relative Humidity, Supersaturation, and Supercooling

Point A (Fig. 5–1) is not on the saturated vapor pressure curve. At 15°C the saturated vapor pressure is 17 mb, whereas at A the pressure is only

---

*In a gas mixture the term *partial pressure* refers to the pressure that one component would have if alone. *Vapor pressure* is the partial pressure of a condensible species. Gases other than the condensible vapor do not affect our discussion.

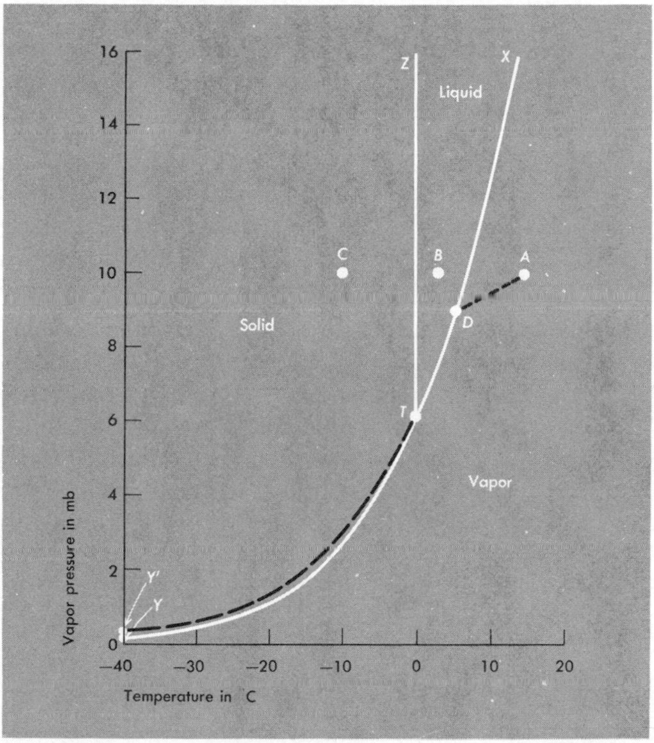

FIGURE 5-1 *Saturated vapor pressure for water and ice. Ice is stable in the region ZTY; liquid in the region ZTX; vapor in the region XTY. Along XT liquid and vapor are in equilibrium. Along YT solid and vapor are in equilibrium. The broken line TY' refers to equilibrium with supercooled water (see text). The points A, B, C, and D are explained in the text.*

10 mb. We say that the *relative humidity* at A is $100 \times 10/17 = 59\%$. When air is saturated, its relative humidity is 100% by definition. Under conditions that are common in the southwestern United States, air can have as low a relative humidity as 10%. The air can then take up nine times as much water as it already contains before it becomes saturated.

Suppose we have water vapor at $+15°C$ and 10 mb partial pressure and chill it to $+3°C$ (point B, Fig. 5–1). In the long run the vapor will change to liquid water, but this may not happen immediately. For a period of time the vapor may exist at a higher vapor pressure than is consistent with saturation. For this to happen no liquid water may be present. If there is any liquid water, molecules from the vapor will immediately attach themselves. In the absence of a liquid surface, however, the water may remain as a vapor. The vapor is then said to be *supersaturated*.

Now consider liquid water at $+3°C$ and 10 mb pressure (point B), and imagine that it cools rapidly to $-10°C$ (point C). In the long run the water must turn to ice, but in the absence of any ice surfaces it may remain liquid for a period of time. The liquid water is then said to be *supercooled*. Because water can be supercooled, we are able to draw the broken curve

Condensation and clouds

TY' for liquid and vapor saturation at temperatures below 0°C. We shall see that cloud droplets often exist in a supercooled state, and that this is an important factor in the development of rain, snow, and hail.

The Ascent of Moist Air

If unsaturated air is cooled sufficiently, its water vapor will condense. There are many ways in which air can be cooled until saturated, but we shall concentrate on the most important process, the cooling of unsaturated air by ascent.

Consider the situation shown in Fig. 5–2, in which a parcel of moist air rises from point *A* to point *T*, without exchanging heat with its surroundings. We shall assume that conditions at ground level are the same as at point *A* in Fig. 5–1, that is, the temperature is +15°C and the water vapor pressure is 10 mb.

Initially, the temperature of the rising air falls at the adiabatic lapse rate, decreasing at 9.8°C km^{-1}. The pressure also falls as the air rises, and

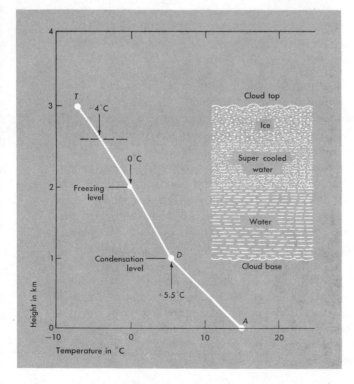

FIGURE 5-2 *Ascent of a parcel of moist air. The trajectory starts at A and ends at T. Condensation starts as soon as saturation is reached, but freezing does not occur at the 0°C level. For the sake of illustration, it is assumed that freezing begins at −4°C.*

Condensation and clouds

the overall effect is for the parcel to follow the dotted line *AD* in Fig. 5–1. At a temperature of +5.5°C and a pressure of 9 mb of water vapor (point *D*), saturation is reached. Since large supersaturations are rare in the atmosphere, this is close to the level at which condensation occurs (the *condensation level*). The condensation level marks the base of the clouds.

In considering further changes of temperature as the air continues to rise we must take account of the *latent heat of condensation* and later the *latent heat of freezing*. If 1 gram of water vapor condenses to liquid at 0°C, it releases 597 calories* or 2.5×10^{10} erg. If 1 gram of liquid freezes at 0°C, 80 calories or 3.3×10^9 erg are released. If water changes directly from vapor to solid, the latent heat released is equal to the sum of these two numbers, 677 calories or 2.8×10^{10} erg.

Let us now reconsider the discussion of the adiabatic lapse rate that appears in the Appendix. There is no mention there of latent heat as a form of energy, but a complete statement of the first law of thermodynamics requires that latent heat be included. When we calculate the heat required to cool a parcel of air in process *a* in Fig. A–1, we use a specific heat of c_p erg gm^{-1} deg^{-1}. If the air is saturated, however, water will condense as the air cools. The latent heat released during the condensation may be enough to substantially increase the total quantity of heat that must be extracted from each gram of air in order to lower the temperature by 1°C. This total quantity of heat can amount to as much as three times c_p for air at ground level saturated at 37°C; at low temperatures, on the other hand, when the vapor pressure is low, the latent heat released may be negligible.

The net effect of condensation, therefore, is to give a higher effective value of c_p and a lower effective value of the adiabatic lapse rate $\Gamma = g/c_p$. In the tropics the temperature of a moist air parcel may decrease with height at a rate as small as $\Gamma/3$ (about 3°C km^{-1}). In the upper troposphere, however, since air is too cold to hold much water vapor, condensation has little influence on temperature. The lapse rate, therefore, is close to Γ, whatever the relative humidity.

In Fig. 5–2 the effect of condensation upon the adiabatic process is shown by the curve *DT*, which has a substantially smaller lapse rate than the portion *AD*. At a height of about 2 km, in this example, the temperature falls to 0°C (the *freezing level*). Usually, water will not freeze right away but will become supercooled instead. At some temperature below 0°C the water changes to ice. We cannot predict with confidence just where this will occur. For purposes of illustration we have marked it at −4°C and 2.5 km.

Finally, on our trajectory we come to point *T*, where for one reason or another, the air ceases to rise. This level marks the top of the clouds.

*One calorie is equal to 4.18×10^7 erg and is the heat required to raise the temperature of 1 gm of water by 1°C.

Condensation and clouds

Motions of the terrestrial atmosphere are such that they produce an irregular and varying pattern of rising and descending air. The distribution of clouds over the surface of the globe is, therefore, irregular and varying as well (see Fig. 5–3). The situation is strikingly different on Venus (see Fig. 5–4), which is completely covered with what appears to be a uniform layer of cloud.

FIGURE 5-3 *The Earth photographed by the Applied Technology Satellite from a height of 36,000 km above Brazil. (Courtesy of National Aeronautics and Space Administration.)*

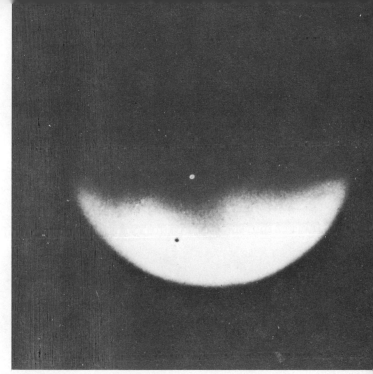

FIGURE 5-4 *Venus. The uneven terminator is caused by the presence of clouds. (Courtesy of Mount Wilson and Palomar Observatories.)*

Condensation on Other Planets

The cloud cover on Venus dominates many aspects of the behavior of the lower atmosphere, as our discussion in previous chapters has shown. Nevertheless, we do not know the nature of the condensate in the outermost, visible layer, let alone whether or not there are more cloud layers below. The only certainty is that the clouds are not solid carbon dioxide, even though this gas forms the bulk of the atmosphere. Figure 5–5 includes the vapor pressure curve for carbon dioxide. The cloud tops of Venus have a temperature close to 240°K and a pressure of about 100 mb. This point is marked on the figure, and it is clear that carbon dioxide can exist only in the form of vapor under these conditions.

On Mars, unlike Venus, clouds are irregularly distributed and ephemeral. Yellow dust clouds have often been recorded, and water mists can probably form close to the ground, perhaps also at cirrus levels. The important feature of the Martian atmosphere, however, is that carbon dioxide can condense out as a solid, certainly at the ground, and possibly also as a high altitude haze.

The pressure of carbon dioxide at the surface of Mars is about 6 mb, and from Fig. 5–5 we see that the gas will condense to a solid at a temperature of about 145°K. This is not an abnormally low temperature on Mars. Even in tropical regions (see Fig. 3–11) the ground temperature can drop

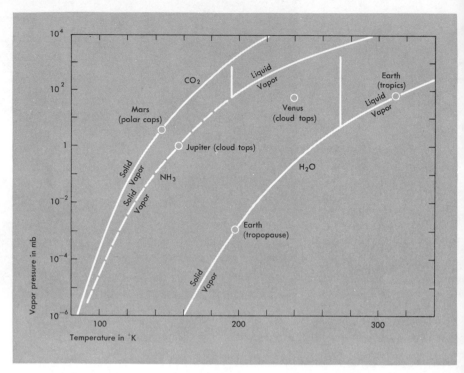

FIGURE 5-5 *Comparison of saturated vapor pressures of H_2O, CO_2, and NH_3 with conditions of temperature and pressure on Mars, Earth, Venus, and Jupiter.*

to 160°K at night. In the polar night the temperature drop is limited only by the latent heat released from the carbon dioxide that rushes in and condenses there. There is no reasonable doubt that the white polar caps on Mars (see Fig. 5–6) consist almost entirely of solid carbon dioxide.

The situation in the polar regions is different from that on Earth because on Mars almost the whole atmosphere could condense if the polar night were long enough, but on Earth only water vapor is condensable, and water vapor is a small fraction of the atmosphere. However, the latent heat that would be liberated if the whole Martian atmosphere were to condense greatly exceeds the total amount of heat that can be radiated away by the polar cap while the Sun is below the horizon. During the winter night about 20% of the atmosphere may condense, and the atmospheric pressure should respond with a rhythmic seasonal change with two minima per year at the two equinoxes, the times when the polar caps are most extensive. This annual variation of pressure on Mars has yet to be measured, but it is confidently anticipated.

The appearance of Jupiter is quite unlike that of any of the inner planets

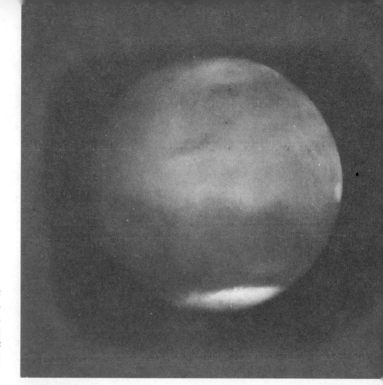

FIGURE 5-6 A *distant view of Mars from an approaching Mariner spacecraft. The polar cap is at the bottom. (Courtesy of National Aeronautics and Space Administration.)*

(see Fig. 5–7). Jupiter is covered with spectacular, changeable clouds, many of which have a yellow or red coloration. We know the total amount of ammonia above the clouds (see Table 1–2) and we can calculate the partial pressure of the gas to be approximately 1 mb at the cloud tops. According to the data available, ammonia at this partial pressure should condense to a solid at 160°K, the approximate temperature of the cloud tops. It is therefore a widely held view that the observed clouds consist of solid ammonia.

We are less sure of the reason for the cloud coloration. The most promising suggestion is that it may have to do with several layers of cloud, with a white ammonia cloud uppermost and with other layers below formed from colored materials. The changeable colors are then associated with rising air motions that give rise to sheets of ammonia cirrus cloud, with sinking regions in between where lower, colored clouds can be seen.

Condensation Nuclei

In our discussion so far we have tended to take for granted the formation of cloud droplets in saturated air. On a microscopic level, however, the situation is not altogether straightforward. Molecules in the vapor phase travel in a random fashion, and we must try to understand how a large number of molecules can come together to form a small, spherical droplet.

FIGURE 5-7 *Images of Jupiter in red light (above) and in blue light (below). (Courtesy of Lunar and Planetary Laboratory, University of Arizona.)*

Our earlier discussion of saturated vapor pressure, culminating in the data presented in Fig. 5-1, referred to a flat interface between vapor on the one hand and solid or liquid on the other. The saturated vapor pressure, however, is not the same over a curved surface, such as that of a cloud droplet. A molecule is less strongly attached to a convex curved surface than to a flat surface. Molecules, therefore, evaporate more readily from such a curved surface. The rate of condensation, on the other hand, is not changed by the shape of the surface. If a balance is to exist between evaporation and condensation, a higher vapor pressure is needed to provide the necessary return flow of molecules to a convex surface.

Thus, the saturated vapor pressure over the surface of a drop is larger than over a flat surface. This excess vapor pressure increases as the drop radius decreases, that is, as the surface becomes more curved. The excess is negligible if the droplet radius exceeds 0.1 μ (1 micron = 10^{-4} cm), but is about 200% if the radius is 0.001 μ.

We now have an interesting problem. If a droplet is to form from the vapor, a small aggregate of individual molecules must first be produced. But a small aggregate will require a large vapor pressure for equilibrium because without this large vapor pressure it will immediately evaporate. A droplet with a radius of 0.01 μ, for example, requires a supersaturation of 12.5% with respect to a flat surface if it is to survive. This is about as great a supersaturation as is observed in the atmosphere, and it is too small to support the existence of any droplet smaller than 0.01 μ. But a 0.01 μ drop contains about 10^5 molecules, and these are most unlikely to come together by accident. We must, therefore, find some way of creating droplets of radius 0.01 μ or greater before clouds can condense.

A way out of this dilemma is provided by small solid particles that are always present in the atmosphere. These are formed by chemical reactions, from products of combustion, from wind blown surface dust, and from ocean spray; they give us the nuclei with radii greater than 0.01 μ that are needed to start the condensation process. Particles acting in this manner are called *condensation nuclei.*

Condensation nuclei are present in large but highly variable numbers. If we assume a typical measured value of 10^3 cm^{-3} for their concentration, we can calculate the approximate size of the water drops that first condense. Suppose that there are N nuclei per cm^3 and that a droplet of radius r cm condenses onto each nucleus. Suppose also that m gm of water condense in each cubic centimeter of air. We may equate m to the product of droplet volume, density of water (ρ), and number density of drops (N)

$$\frac{4}{3} \pi r^3 N \rho = m \tag{5-1}$$

A typical value of m in a condensing cloud is 10^{-6} gm cm^{-3} (or 1 gm per cubic meter); N we are taking as 10^3 cm^{-3}, and $\rho = 1$ for water. Hence

$$r = \sqrt[3]{\frac{3 \times 10^{-6}}{4 \times \pi \times 10^3 \times 1}} = 0.7 \times 10^{-3} \text{ cm} = 7 \text{ μ} \tag{5-2}$$

Growth of Cloud Droplets

What happens to the cloud drops once they have formed depends on the velocity of the vertical wind within the cloud. We must distinguish between the two most important kinds of cloud, *cumulus* and *stratus.*

Condensation and clouds

Cumulus clouds are associated with isolated upward currents of warm air from heated layers near the ground. They are relatively small in lateral extent (1 to 10 km is common) and can rise as high as 10 or 15 km. The vertical wind within a cumulus cloud is usually at least 1 m sec^{-1} and may be as great as 30 m sec^{-1}. Stratus clouds occur in air that is rising much more slowly (at about 10 cm sec^{-1}) in weather systems that may have lateral extents as large as 1000 km. They are found at all levels in the troposphere.

Cloud drops do not retain their original radii (about 7 μ) for long. There is a continual tendency for large drops to grow larger and for small drops to disappear. As an example, the drop sizes in a young cumulus cloud are shown in Fig. 5–8. A cloud that has existed for a long time has an even greater proportion of large drops, some as large as 50 μ radius.

The size relationship between condensation nuclei, "typical" cloud drops, and "large" cloud drops is shown schematically in Fig. 5–9. Since the volume of a drop varies as the cube of the radius, it requires 125 "typical" drops of 10 μ radius to make one "large" drop of 50 μ radius.

In a cloud of water droplets the most important mechanism causing differential growth is *coalescence*, which occurs because drops of different size fall downward at different speeds. The steady fall speed that a drop attains after a period of time under the action of gravity is called the *terminal velocity*, and its value is indicated in Fig. 5–9 against each of the drops shown. For a given material, the larger the drop the greater the terminal velocity; water drops with radii less than about 50 μ follow *Stokes' law*, which predicts that terminal velocity varies as the square of the drop

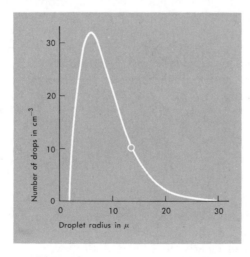

FIGURE 5-8 *A typical distribution of cloud droplet sizes in a small cumulus cloud. The curve shows the number of droplets per cubic centimeter whose sizes lie in a 1 μ range. Thus, the point plotted at 13 μ indicates that 10 drops cm^{-3} have radii between 12.5 μ and 13.5 μ. (From B. J. Mason, 1962, Clouds, Rain and Rainmaking, Cambridge University Press.)*

Condensation and clouds

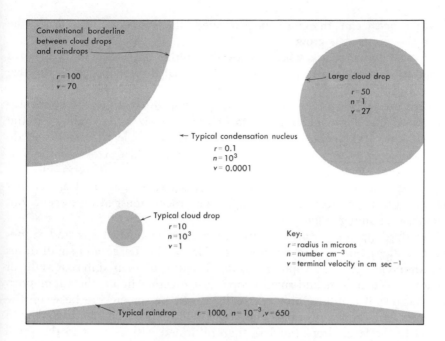

FIGURE 5-9 *A comparison of the sizes, concentrations, and terminal falling velocities of some of the particles involved in condensation and precipitation processes. (From J. E. McDonald, 1958, Advances in Geophysics, Academic Press.)*

radius. A typical condensation nucleus has a terminal velocity of 10^{-4} cm sec^{-1}, and for most practical purposes it can be considered not to settle at all. A typical cloud drop has a terminal velocity of 1 cm sec^{-1}. It takes a day to fall through a cloud 1 km thick and is also, for practical purposes, in permanent suspension. A very large drop, of 100 μ radius, falls at 70 cm sec^{-1}, or 2.6 km each hour; its lifetime before it falls to the bottom of a cloud is not likely to exceed one hour.

The way in which the coalescence mechanism operates is shown in Fig. 5–10(a) for still air. A large drop overtakes all the small drops shown between the broken lines; it is potentially capable of sweeping up these drops. Many details of the process have to be considered to determine the exact rate of growth of large drops, but we can see that the direction of change is towards a few larger drops at the expense of many smaller drops.

Figure 5–10(b) shows the same situation, but with an updraft, such as exists in almost every cloud. The same sweeping-up process can take place, but the large drop will now take longer to fall through the cloud. The growth process can, therefore, be extended; the stronger the updraft, the larger the drops can grow.

We now have to ask what causes the initial difference in drop sizes needed to start the coalescence process and what in practice limits the size to which drops can grow by coalescence.

There are three answers to the first question. First, different condensation nuclei can form initial droplets of different sizes, depending on the size and chemical composition of the nuclei. If the nuclei are small and do not dissolve in water, we expect a rather uniform size, close to the 7 μ radius that we calculated. Large salt nuclei, however, can form substantially larger drops. A second process involves coalescence of droplets that touch as a result of accidental jostling. This process must always occur, but it may be of minor importance.

The third process involves evaporation from smaller drops and subsequent condensation on larger drops (see Fig. 5–11). Large and small drops have different saturated vapor pressures because of their different radii of curvature. When large and small drops exist together in air, the vapor pressure, because it can have only one value, must be somewhere between the two saturated vapor pressures. The vapor will then be supersaturated with respect to the large drops but less than saturated with respect to the small

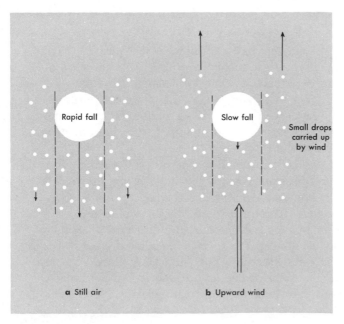

FIGURE 5-10 *The coalescence process. The small drops that might be swept up by the large drop are contained between the broken lines. (a) In still air, shown on the left, the process can continue only until the large drop falls out of the bottom of the cloud. (b) If there is an updraft, as shown on the right, it takes longer for the drop to fall out of the cloud and it therefore has a longer time in which to grow.*

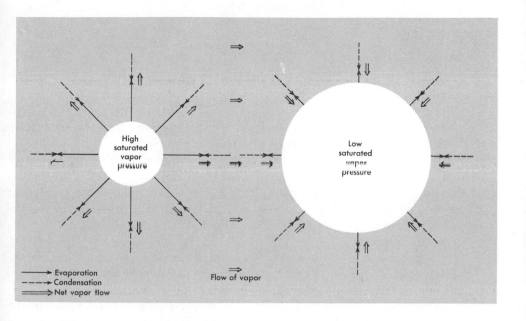

FIGURE 5-11   *Droplet growth by evaporation and condensation.*

drops. Consequently, water evaporates from the small drops while it condenses onto the large drops. The net result is that large drops grow at the expense of small drops.

Now let us consider the maximum size to which a drop can grow by the coalescence process. The maximum drop size is clearly related to the time a large drop spends in the cloud and the number of smaller drops available to be swept up. The time a large drop can spend in a mature cloud is the time it takes to fall to the bottom. This depends on the vertical extent of the cloud and the magnitude of the updraft, which can slow the rate of descent of the drop. The thicker the cloud and the greater the updraft, the larger the droplets can grow.

Consider a low-level stratus cloud that contains only water droplets. It will typically be less than 1 km thick and will have vertical velocities less than 10 cm sec^{-1}. Under these conditions the maximum drop radius will be less than 100 μ. An observer at the base of such a cloud will note the existence of a damp mist, with some droplets falling at about 70 cm sec^{-1}. If there is a clear layer of unsaturated air below the cloud, such droplets will probably evaporate before they reach the ground.

Condensation and clouds

Rain from Warm Clouds

Let us now see how the discussion of the previous section must be modified if the vertical velocity rises to 10 cm sec^{-1} or greater.

According to one calculation, a stratus cloud containing 400 cm^{-3} of 8 μ liquid water drops, with an updraft of 10 cm sec^{-1}, will, by the coalescence process, produce drops up to 150 μ in radius. Such drops can fall at more than 1 m sec^{-1} and survive a short period in a dry layer below the cloud. These drops could reach the ground in the form of *drizzle*, the lightest form of rain.

In tropical regions, however, most of the clouds are cumulus clouds. Since much of the world's total rainfall occurs in the tropics, the mechanism of rain from warm cumulus clouds is one of great importance.

The crucial differences between a cumulus and a stratus cloud are the vertical extent of the cloud and the magnitude of the vertical updraft; the updraft can be several meters per second in a cumulus cloud. In order to assess the importance of such updrafts, let us consider the specific example of a model of a cloud whose basal temperature is 20°C. This is warm for a cloud base and the cloud will contain a large amount of liquid water. Suppose that the updraft is 3 m sec^{-1} and that we start with a drop of radius 20 μ. The terminal velocity of the drop is 4 cm sec^{-1}; it will, therefore, rise with a velocity of 296 cm sec^{-1}. As it rises, it will grow until it has a radius of about 400 μ, when its terminal velocity equals the updraft velocity. After this the drop will begin to fall. The apex of the trajectory is approximately 2.4 km above the base. The drop will continue to grow as it descends and it will leave the base of the cloud as a large drop, approximately 2500 μ (2.5 mm) in radius (see Fig. 5–12).

Drops of 2.5 mm radius are typical of those that appear at the beginning of a rain shower. We need an additional process to account for the rapid rate at which these drops appear once a shower is underway. This process depends on the instability of drops with radii greater than 2.5 mm. Large drops tend to break up into a few large fragments and many smaller droplets. The larger fragments rise again in the updraft and grow to an unstable size in a few minutes. They, in turn, fracture and produce more large fragments. A chain reaction is now underway, which proceeds with rapidity until the cloud has lost a substantial amount of its liquid water.

The processes of shower and drizzle formation are illustrated schematically on the right and left sides, respectively, in Fig. 5–13. We must now

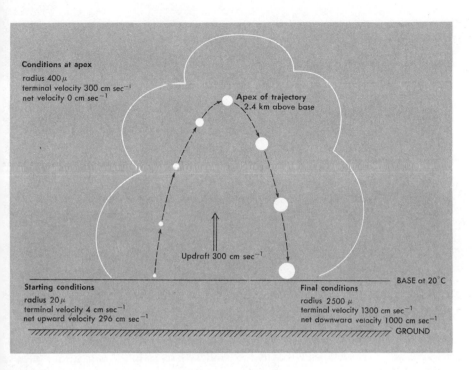

Conditions at apex
radius 400 μ
terminal velocity 300 cm sec⁻¹
net velocity 0 cm sec⁻¹

Apex of trajectory
2.4 km above base

Updraft 300 cm sec⁻¹

BASE at 20°C

Starting conditions
radius 20 μ
terminal velocity 4 cm sec⁻¹
net upward velocity 296 cm sec⁻¹

Final conditions
radius 2500 μ
terminal velocity 1300 cm sec⁻¹
net downward velocity 1000 cm sec⁻¹

GROUND

FIGURE 5-12   *A numerical model of the production of a large raindrop by coalescence in a warm cloud. A cloud drop at the base has a small terminal velocity and is carried upward, continually growing, until its terminal velocity is equal to the updraft velocity. Thereafter it falls, finally reaching the ground as a raindrop.*

consider the effects that arise if part of the cloud is above the freezing level and contains supercooled water.

Snow, Hail, and Rain from Freezing Clouds

A region containing supercooled water always has a small number of ice crystals. These exist because occasional condensation nuclei have the property of initiating freezing. The number of nuclei that can do this is very small indeed for temperatures above −10°C, but the number of these freezing nuclei increases as the temperature decreases. At about −33°C water begins to freeze spontaneously without the help of foreign matter. By

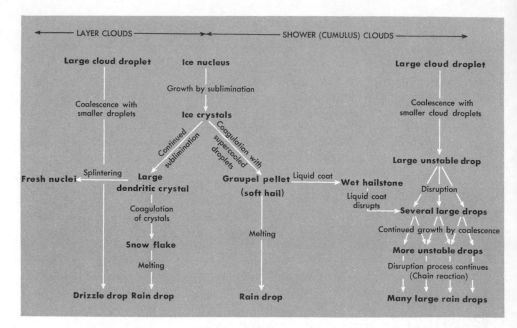

FIGURE 5-13 *Natural precipitation mechanisms. (From B. J. Mason, 1957, Physics of Clouds, The Clarendon Press, Oxford.)*

$-40°C$, the spontaneous process is so rapid that the purest water freezes immediately.

In the atmosphere we are concerned with water supercooled to about $-10°C$. At this temperature a typical concentration of freezing nuclei is one per cubic meter or 10^{-6} cm^{-3}. Under these conditions there are about one billion water droplets for every ice crystal.

Now consider the saturated vapor pressures over the ice and supercooled water, as shown in Fig. 5-1. At $-10°C$ the vapor pressure over supercooled water is 2.9 mb; over ice it is 2.6 mb, a difference of 10%. We therefore have an analogous situation to that shown in Fig. 5-11, except that the ice crystal replaces the large drop in having the lower saturated vapor pressure. Distillation occurs from water drops to ice (the term *sublimation* is used to describe the vapor-to-solid transition), and the ice crystal grows.

An important point is that the growth is much more rapid here than in distillation from one water drop to another. If we consider a mixture of 10 μ and 1 μ water drops, the vapor pressure difference is only 0.11%; for the sublimation process at $-10°C$, on the other hand, the difference is 100 times greater, and the rate of growth of the ice crystal is that much more rapid.

If sublimation continues for a period of time, a complicated dendritic crystal develops—the familiar snowflake. It may fall to the ground, but a snowstorm will not develop unless the rare ice crystal, which started the process, can be multiplied. The mechanism of multiplication is *splintering*. Parts of the snowflake drop off, forming the small ice crystals required for a new sublimation process. Again a chain reaction is established. The events are illustrated in the left-hand center column of Fig. 5–13, with the additional feature that the snowflake is shown to melt and change to a raindrop.

Finally, the right-hand side of the center column in Fig. 5–13 illustrates a mixed sublimation-coalescence process leading to hail stones. Once a crystal has grown to a substantial size by sublimation, it can fall through the cloud and coalesce with supercooled water droplets, collecting layer after layer of ice. A solid hailstone does not become unstable as does a water drop greater than 2.5 mm radius. There is, therefore, no comparable limit to the size to which a hailstone can grow. The largest hailstone reported in the U.S. had a diameter of 13.8 cm and weighed close to 700 gm.

6

The evolution of atmospheres

We have shown how the behavior of an atmosphere depends on its composition and its mass. It is now time to consider the processes governing the composition and mass of an atmosphere and to describe how the atmospheres of the solar system could have developed the properties that they have today. This aspect of atmospheric science has many unanswered questions, and it is possible that the theories we shall describe will in the future turn out to be wrong. However, we cannot understand the workings of the world we live in without understanding its history. Some account of this history is therefore necessary, even if the account is highly speculative.

We must first realise that our atmosphere is not an isolated and static layer of gas. Various chemical constituents are continually being added to the atmosphere while other constituents are continually being removed. One example is carbon dioxide.

The Cycles of Carbon Dioxide

The ocean is the largest source and also the largest sink for carbon dioxide in the Earth's atmosphere. Atmospheric carbon dioxide dissolves in the ocean and is then converted to bicarbonate ion. At the same time, oceanic bicarbonate ion is converted by the reverse process into carbon dioxide, which subsequently escapes to the atmosphere. This cycle, which converts atmospheric carbon dioxide to oceanic bicarbonate ion and back again, is shown on the right in Fig. 6–1. Values are given for the rate at which the gas dissolves and for the rate at which the gas comes out of solution. The two processes occur about equally fast, which indicates that

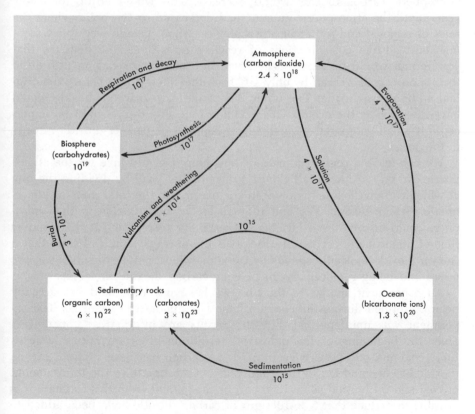

FIGURE 6-1 *The cycles of carbon and carbon dioxide. Boxes denote reservoirs of carbon; the contents of the reservoirs are expressed in gm of CO_2. The arrows denote the transfer of material between reservoirs; the rates of transfer are given in gm of CO_2 per year.*

The evolution of atmospheres

an equilibrium exists between the pressure of carbon dioxide in the atmosphere and the concentration of carbonate ion in the sea. It is principally this equilibrium that governs the amount of carbon dioxide in the atmosphere.

Suppose we were to add 2.4×10^{18} gm of carbon dioxide to the atmosphere by artificial means. The data in Fig. 6–1 show that this amount is equal to the total mass of carbon dioxide in the atmosphere today. Initially, therefore, the amount of carbon dioxide in the atmosphere would be doubled by our addition, but this would not last for long. The rate at which carbon dioxide dissolves in the ocean is proportional to the pressure of carbon dioxide in the atmosphere. This rate would now be doubled, whereas the rate of return of the gas to the atmosphere would not be changed. Since the rate of loss of carbon dioxide to the oceans would exceed the rate at which the oceans release carbon dioxide to the atmosphere, the amount of atmospheric carbon dioxide would decrease. This would go on for a few years, until the atmospheric amount had decreased to the point where the rates of removal and return of atmospheric carbon dioxide were once more in balance. From considerations of reaction rates it can be deduced that equilibrium would be reestablished with the amount of carbon dioxide in the atmosphere increased by only 10% from the value it had before the original addition was made. The other 90% of the extra carbon dioxide would have dissolved in the ocean. This addition (see Fig. 6–1) would be imperceptibly small compared with the large amount of carbon dioxide already in the ocean.

We can make a rough estimate of how long it would take for the atmospheric carbon dioxide to fall almost to its original level by dividing the rate at which carbon dioxide dissolves in the ocean into the total mass of atmospheric carbon dioxide. We find that all the carbon dioxide in the atmosphere could dissolve in as little as six years. Six years, then, is the turnover time of atmospheric carbon dioxide. It is the average length of time that a carbon dioxide molecule spends in the atmosphere before dissolving in the ocean and being replaced by an oceanic carbon dioxide molecule.

An experiment similar to the one we have just described has actually been performed during the course of the last hundred years or so. The experiment was unintentional, but the results are informative nonetheless. Since the beginning of the industrial revolution great quantities of fossil fuel (coal and oil) have been burned. As carbon has been burned, carbon dioxide has been added to the atmosphere. An estimate of the total amount of coal and oil burned since 1860 can be made, and from this estimate it is possible to deduce that 5×10^{17} gm of carbon dioxide have been added to the atmosphere in the same period. Actual measurements of the carbon dioxide content of the atmosphere show a smaller increase, however. A part of the added carbon dioxide has dissolved in the ocean.

The burning of fossil fuel is an element of another cycle that affects atmospheric carbon dioxide; this is the cycle of photosynthesis and respiration, which is shown on the left in Fig. 6–1. Plants extract carbon dioxide from the atmosphere or ocean and, with the addition of water, manufacture carbohydrates and release oxygen. The reverse reaction occurs as well. Plants and also animals consume atmospheric oxygen and burn their carbohydrates, producing carbon dioxide and water. This process, known as *respiration*, furnishes the organism with energy. Another process that restores to the atmosphere the carbon that has been stored in animals and plants is decay, which occurs after the death of the organism.

The rate at which respiration and decay occur is very nearly equal to the rate of photosynthesis, as the values in Fig. 6–1 show. Nevertheless, there is, on average, a small imbalance that is important. Some of the organic debris is washed out to sea and is buried by sand and mud before it has decayed. As a result, the rate of photosynthesis is slightly faster than the rate of respiration and decay, and there is a continual addition of carbon of organic origin to the sediments at the bottom of the sea. In time these sediments are converted to rocks, and the carbon is incorporated therein, but the process does not stop there. The layer of sedimentary rocks on Earth is not getting thicker all the time. Instead, the rocks are lifted up above the surface of the sea, where they are subjected to weathering and erosion. The carbon in the rocks is oxidized in the process and returned to the atmosphere as carbon dioxide. In this respect, the burning of fossil fuels may be thought of as a greatly accelerated form of weathering. At the present time, the burning of fossil fuel is producing carbon dioxide about thirty times as fast as rock weathering.

Not all the carbon in sedimentary rocks is in the reduced form (carbon without oxygen) that we have been discussing. About 85% of the sedimentary carbon occurs in the form of carbonate rocks such as limestone ($CaCO_3$). When these rocks are weathered the limestone dissolves, and bicarbonate ions, together with calcium ions, are washed down to the sea. Clearly these ions cannot simply accumulate in the sea. We can see from Fig. 6–1 that carbonate ions are being added as a result of weathering at a rate that is large enough to supply all of the carbonate in the sea in about 100,000 years—a time span much shorter than the age of the Earth. What happens is that the carbonate ions and the calcium ions are deposited on the floor of the ocean as sediments, sediments that are eventually exposed to the atmosphere and weathered. Some of the carbon is returned to the atmosphere through volcanic action. So here again we have a cycle, in this case a cycle of weathering and sedimentation, that links rocks and ocean. By dividing the rate of weathering into the carbonate content of the rocks, we can calculate the average lifetime of the rocks. The values in Fig. 6–1 give a lifetime of 300 million years.

For simplicity we have described the various cycles of carbon as though they proceed always at the same rate. In fact, there is no reason to suppose

The evolution of atmospheres.

that this is so. It must be remembered, however, that the rates of many of the processes involved in the various cycles of carbon can vary without changing the amount of carbon dioxide in the atmosphere. We have described how this amount is largely determined by equilibrium with the sea. This equilibrium depends on factors such as the average temperature of sea water, but it does not depend directly on the rate of volcanic production of carbon dioxide, for example, or on the rate of burial of organic carbon.

Carbon Dioxide on Venus

It is possible that the amount of carbon dioxide in the atmosphere of Venus is also determined by equilibrium, but equilibrium with rocks rather than with an ocean. As an example of the kind of process that can affect the composition of the atmosphere of Venus, let us consider the reaction of carbon dioxide with the mineral wollastonite (calcium silicate). If a quantity of carbon dioxide were added to the atmosphere of Venus, by volcanoes for example, the amount of carbon dioxide would not show a long-term increase. Instead, the additional carbon dioxide would react with wollastonite, which we assume to be present at the surface of the planet, producing two different minerals, calcite (calcium carbonate) and quartz (silicon dioxide). If, on the other hand, a quantity of carbon dioxide were removed from the atmosphere, the reverse reaction would occur; calcite in the rocks of Venus would react with quartz, releasing enough carbon dioxide to restore the atmosphere to its original level.

Study of this hypothetical equilibrium between atmospheric gases and rocks on Venus indicates that the reaction with wollastonite is only one of a number of reactions that may control the amount of carbon dioxide in the atmosphere. We cannot be sure just which reactions are the most important until we know more than we do about the properties of the surface of Venus. But uncertainty concerning the detailed reactions does not detract from the conclusion that the pressure of carbon dioxide in the atmosphere of Venus is probably in equilibrium with the rocks.

Examination of chemical reactions that could affect the other gases in the atmosphere of Venus shows that every gas in the atmosphere could be in equilibrium with rocks rather similar in composition to those on Earth. This remarkable result may mean that the chemical composition of the atmosphere of Venus is entirely determined by the properties of the surface. The situation is quite different on Earth where the major gases, oxygen and nitrogen, as well as most of the minor gases are far from being in chemical equilibrium with the ground. Why are Earth and Venus so different in this regard?

There are two factors that must be considered in answering this question. First is the existence of processes tending to drive the atmosphere away from equilibrium, and second is the rate of the chemical reactions tending to restore the atmosphere to equilibrium. Photosynthesis is a good example of a disequilibrating process. In the absence of photosynthesis, atmospheric oxygen would recombine with the organic carbon in rocks, producing carbon dioxide. In equilibrium, there would be no oxygen in the atmosphere. Photosynthesis serves to separate carbon and oxygen, tying the carbon up in living creatures and their remains, and releasing the oxygen to the atmosphere. So, one important difference between Earth and Venus is that Earth has photosynthesizing life, but Venus almost certainly does not. Surface temperatures on Venus are too high to permit life.

The high surface temperature is also responsible for the other factor causing the atmosphere of Venus to be close to chemical equilibrium. At high temperatures, the chemical reactions between atmospheric gases and rocks occur rapidly. Thus, the approach to equilibrium is fast. At terrestrial temperatures, on the other hand, reactions are so slow that most of the gases in the atmosphere never get close to equilibrium.

Why did a high surface temperature develop on Venus? In order to answer this question, we must consider some of the facts and theories concerning the origin and evolution of the solar system. We believe that the planets and the Sun all condensed at about the same time, 4.6 billion years ago, from a great cloud of gas and dust. We can put together a fairly complete picture of the chemical composition of the primordial cloud, partly by spectroscopic analysis of sunlight, and partly by chemical analysis of meteorites and of terrestrial materials.

When we compare the primordial composition with the composition of the Earth, we find that there are striking similarities and also striking differences. In particular, there are a number of elements that are relatively as abundant on Earth as they are in the solar system. These are all elements that take part in chemical reactions and were probably present as solid or liquid compounds in the primordial cloud. On the other hand, the family of inert gases—helium, neon, argon, krypton, and xenon—which hardly react at all and could have existed only as gases, is conspicuously absent. The Earth has 10^6 times less xenon relative to silicon, for instance, than the solar system as a whole.

The implication is either that gases were not incorporated in the Earth when it was formed or else that the entire atmosphere was lost subsequent to formation. In either case there must have been a time when the primitive Earth had no atmosphere. Presumably the same was true of the other inner planets. The very important conclusion is that the atmospheres that exist today on Earth, Mars, and Venus are not remnants of the original solar cloud. Instead, they are the result of the release of gases from the minerals that make up the solid parts of these planets.

The evolution of atmospheres

Lacking definite information on the chemical composition of the gases released to the atmospheres of Mars and Venus, let us assume that Earth is representative. On Earth the gas that has been released to the atmosphere in the greatest abundance is water vapor, with carbon dioxide the next most abundant. We can see this by comparing the mass of water at the surface of the Earth (1.6×10^{24} gm) with the mass of carbon dioxide that has passed through the atmosphere and is now tied up in sedimentary rocks (4×10^{23} gm). Let us now imagine how atmospheres may have evolved on Venus, Earth, and Mars. What would happen if we were to start with planets that had no atmospheres at all and if we were to release from the solid parts of the planets a mixture of water vapor and carbon dioxide?

The Runaway Greenhouse Effect

We first need to decide what the surface temperatures of the planets would be when they had no atmospheres. These temperatures can be determined by equating the infrared radiation emitted by each planet to the solar radiation absorbed by the planet. The calculation was described in Chapter 3; Equation (3–4) presents an expression for the temperature in terms of the flux of solar radiation and the albedo. We do not know what albedos the primitive planets had, but let us assume that they had the same albedo as Mars has today. The values we obtain for the surface temperatures of the primitive planets are shown on the left-hand side in Fig. 6–2. Venus is the hottest because it is the closest to the Sun; Mars is the coldest.

The question we must now consider is how the surface temperatures increase, as a result of the greenhouse effect, when water vapor and carbon dioxide are released to the atmosphere. The greenhouse effect was also discussed in Chapter 3; Equation (3–10) gives the value of the ground temperature in terms of the effective temperature, which we have just evaluated, and the optical thickness of the atmosphere. We shall assume that in this early era water was sufficiently abundant to dominate the absorption and emission of infrared radiation by the atmospheres of Venus, Earth, and Mars. It is then possible to relate the optical thickness of the atmosphere approximately to the vapor pressure of water at the surface, in this way converting Eq. (3–10) to an expression for surface temperature as a function of surface vapor pressure. With this expression we can calculate the way the surface temperatures of Venus, Earth, and Mars evolved as water and carbon dioxide gradually accumulated in their atmospheres.

The evolution is shown in Fig. 6–2, where surface pressure is on the horizontal axis and surface temperature is on the vertical axis. As time

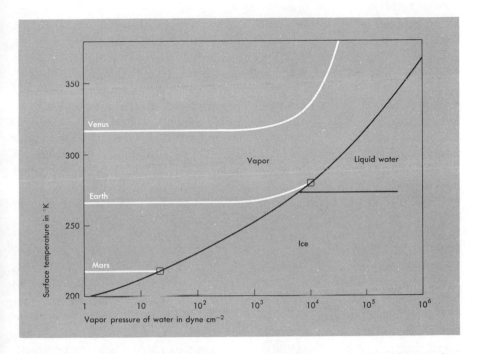

FIGURE 6-2 *The runaway greenhouse effect. The light curves show how surface temperatures increase, due to the greenhouse effect, as water vapor accumulates in the atmospheres of the inner planets. On Mars and on Earth the increase is halted when the water vapor pressure is equal to the saturated vapor pressure (shown as the dark curve), and freezing or condensation occurs. Temperatures are higher on Venus because Venus is closer to the Sun and saturation is never achieved. The temperature runs away. Note that the temperatures on the left-hand axis are not the same for Earth and Venus as the effective temperatures in Table 3-1. They differ because a different albedo has been used. (After S. I. Rasool and C. DeBergh, 1970.)*

passes, and gases accumulate in the atmosphere, we move across the figure from the left-hand side toward the right. The figure also shows the saturated vapor pressure curve of water, which was presented in Fig. 5-1. We must consider how the water in our evolving planetary atmospheres is affected by condensation.

On Mars we find that the temperature is so low that gases do not accumulate in the atmosphere for long before the pressure reaches the saturated vapor pressure of ice. Because the pressure of water vapor in the atmosphere cannot exceed the saturated vapor pressure, any additional water released to the atmosphere condenses on the surface in the form of frost, and the surface temperature ceases to rise. On Mars, therefore, we have a small greenhouse effect associated with a low pressure of water vapor in the atmosphere.

Earth is warmer than Mars because Earth is closer to the Sun. For this reason, ice does not form at the average temperature of the ground. Nevertheless, a point is reached in the evolution of the atmosphere where the pressure is equal to the saturated vapor pressure, and water condenses to form oceans. From this point on, the situations on Earth and Mars are similar. Release of additional water to the atmosphere does not lead to an increase in surface pressure. The additional water condenses. Moreover, in the absence of increases in water vapor pressure, the greenhouse effect cannot produce further substantial increases in ground temperature.

While the growth of surface pressure and temperature on Mars and on Earth is arrested either by freezing or by condensation of water vapor, no similar end to the growth occurs on Venus. Because Venus is closer to the Sun than Earth or Mars, it is warmer, as Fig. 6–2 shows. The saturated vapor pressure of water increases as the temperature increases, which means that a greater pressure of water is required on Venus in order to cause condensation than is required on Earth or Mars. But it turns out that the greenhouse effect caused by the greater pressure of water is so large that condensation never occurs. As more water is released to the atmosphere, the surface pressure continues to increase, and because of the greenhouse effect, the surface temperature also continues to increase.

This behavior of Venus' atmosphere has been called the *runaway greenhouse effect*. There are uncertainties in the theory particularly as regards the albedo of the primitive Venus and the nature of the gases released to the atmosphere from the solid portion of the planet. The account of the evolution of Venus' atmosphere that we are giving is speculative, as are all theories in this field. Nevertheless, the runaway greenhouse theory provides a satisfying explanation of how atmospheric evolution on Earth and Venus happened to diverge in such a striking way.

The idea is that the growth of the Earth's atmosphere was halted at moderate pressure and temperature by the formation of the oceans. Water condensed, and carbon dioxide dissolved in the water, furnishing a suitable environment for the onset of life. On Venus, however, the runaway greenhouse effect prevented oceans from ever forming. Surface temperatures rose so high that atmospheric gases could react rapidly with surface rocks, leading to the chemical equilibrium between atmosphere and surface that we have already described. In the course of these chemical reactions enough condensible material accumulated in the atmosphere to form the extensive clouds that cover Venus today.

According to this theory, there were at one time large quantities of water in the atmosphere of Venus. Today the atmosphere is very dry, as we related in Chapter 1. Is there some process that could have removed the water from Venus' atmosphere?

Escape of Atmospheric Gases to Space

There is such a process. Both Venus and Earth are losing water all the time, although not very fast. Water vapor (H_2O) in the upper atmospheres of the planets is photodissociated by solar ultraviolet radiation into hydrogen atoms (H) and hydroxyl radicals (OH). These species are involved in a number of chemical reactions that ultimately lead to reformation of the water vapor. Not all of the hydrogen atoms recombine, however. Some of them diffuse upward into the thermosphere (see Fig. 3–1) and, even higher, into the region of the atmosphere known as the *exosphere*.

The exosphere is the region at the top of the atmosphere where collisions between molecules are very infrequent. The density of the atmosphere decreases steadily as the altitude increases, according to the barometric law described in Chapter 1. The distance that a given atmospheric molecule can travel before colliding with another molecule gets longer as the density decreases and there are fewer molecules with which to collide. Eventually this distance, the *mean free path*, becomes so long that a molecule travelling upward can go all the way out to space without colliding at all. The altitude at which the mean free path first becomes long enough for escape to space to be possible is known as the *critical level*. The atmosphere above the critical level is the exosphere. On Earth the critical level is at a height of about 500 km, but the height increases and decreases in response to changes in upper atmosphere density. Densities high in the atmosphere of Venus are less than on Earth, and the critical level occurs at a height of about 200 km.

Although atoms in the exosphere of a planet can travel a long way without colliding, they will not escape from the planet unless they are travelling sufficiently fast. The force of gravity pulls most atoms back down to the critical level just as it pulls a baseball or rocket back down to the ground (see Fig. 6–3). In order to escape from the gravitational attraction of a planet, any body must have more than a certain minimum velocity, called the *escape velocity*. We can derive the escape velocity from the law of conservation of energy.

Consider a body of mass m gm travelling at a velocity v cm sec^{-1}. Its kinetic energy is $\frac{1}{2}mv^2$ erg. In order to escape from the Earth's gravitational field, work must be done against the force of gravity. If energy is to be conserved, we must draw upon the kinetic energy to provide this work. Thus, if the particle is to escape, the kinetic energy must exceed the work required to take the body far from the planet.

The work required for a body of mass m gm originally at the surface of a planet of radius R cm and gravitational acceleration g cm sec^{-2} is mgR

The evolution of atmospheres

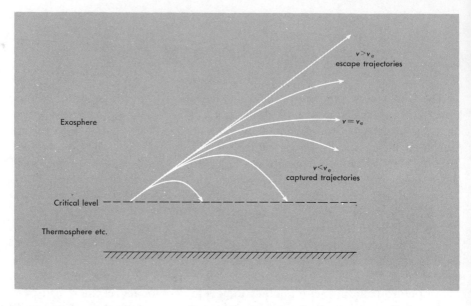

v>v_e escape trajectories

Exosphere

v = v_e

v<v_e captured trajectories

Critical level

Thermosphere etc.

FIGURE 6-3 *Only the atoms whose velocity exceeds the escape velocity v_e can escape from the planet. This may be a minute fraction of the total number of atoms at the critical level.*

erg. We can understand this result by means of the argument that follows. At the surface of the planet the force on the body is mg dyne. Work is equal to force times distance. Thus, to move the body a distance of one centimeter from the surface requires an amount of energy equal to $mg \times 1$ erg. As we go farther from the center of the planet the force of gravity decreases; it follows the inverse square law as does the intensity of solar radiation (see Chapter 1). Thus, the second centimeter requires less work than the first, and so on. When the body is at a distance R from the surface, it is at a distance $2R$ from the center, and the work per centimeter of displacement has fallen by $2^2 = 4$. We may obtain a reasonable estimate of the work required to take the body far from the planet by ignoring gravity for distances from the surface greater than R, while assuming that gravity is constant for smaller distances. By this means we arrive at $mg \times R$ erg as an estimate of the energy that a body must have in order to escape from the planet. The escape velocity v_e is the velocity of a body that has this minimum amount of kinetic energy,

$$\tfrac{1}{2}\, mv_e^2 = mgR \tag{6-1}$$

or

$$v_e = \sqrt{2gR} \tag{6-2}$$

The evolution of atmospheres

Note that the value of the escape velocity that we have derived does not depend on the mass of the body; it is the same for an atom or for a space ship. Calculated values of escape velocity are shown in Table 6–1. The differences from planet to planet may not seem large, but the number of atoms able to escape from an atmosphere depends in a critical way on the relative magnitudes of the escape volcity v_e and the most probable atomic velocity v_0.

If we could measure the velocities of a sufficient number of gas atoms, we would find some atoms with any chosen velocity, no matter how small or how large. The number of atoms with extreme velocities, however, would be small. For the most part atoms tend to have velocities close to the most probable velocity

$$v_0 = \sqrt{\frac{2kT}{m}} \tag{6–3}$$

where k is Boltzmann's constant (1.38×10^{-16} erg deg^{-1}), T is the temperature, and m is the atomic mass. Values of v_0 for a range of temperatures are shown in Table 6–2.

Let us now consider escape of gases from the moon. The escape velocity is 2.3 km sec^{-1}. From Table 6–2 we see that the most probable velocity for hydrogen exceeds this figure for all temperatures above 300°K. The day side of the moon is as hot as this, so most hydrogen atoms on the day side of the moon have enough kinetic energy to escape. The moon's gravitational field is clearly too weak to retain an atmosphere of hydrogen.

Since every species has a few atoms with velocities in excess of the escape velocity, all gases on all planets leak more or less rapidly to space. Thus, if a species is to remain on a planet for a long period of time, it must

Table 6–1

Escape Velocities

Planet	Gravitational Acceleration (cm sec⁻²)	Radius (km)	Escape Velocity (V_e) (km sec⁻¹)
Mercury	376	2,439	4.3
Venus	888	6,049	10.3
Earth	981	6,371	11.2
Moon	162	1,738	2.3
Mars	373	3,390	5.0
Jupiter	2,620	69,500	60
Saturn	1,120	58,100	36
Uranus	975	24,500	22
Neptune	1,134	24,600	24

The evolution of atmospheres

Table 6–2

$$\text{Most Probable Velocities } v_0 = \sqrt{\frac{2kT}{m}} \text{ in km sec}^{-1}$$

Atom	Atomic Weight (amu)	Temperature (°K) 300	600	900
H	1	2.24	3.16	3.87
He	4	1.12	1.58	1.94
O	16	0.56	0.79	0.97

be because the proportion of atoms with velocities exceeding v_e is very small.

Figure 6–4 gives the proportion of atoms whose velocities exceed a given velocity v in terms of v/v_0. We see that for $v = v_0$ the proportion is close to ½ as we might have anticipated. But for $v/v_0 = 4$, we find that only one atom in every 10^6 travels faster than v. Now take the case of oxygen on Earth at 600°K. The escape velocity v_e is 11.2 km sec^{-1} while the most probable velocity v_0 is 0.79. Figure 6–4 shows that, under this circumstance, only 1 atom in every 10^{84} has a velocity larger than the escape velocity. The fraction is so small that the leakage of atomic oxygen from the Earth is

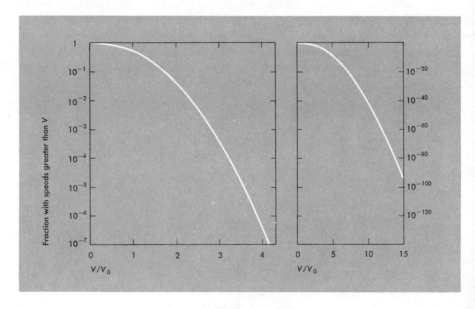

FIGURE 6-4 *The fraction of atmospheric molecules with speeds greater than v in terms of the ratio of v to the most probable velocity v_o. The two parts of the figure are the same, apart from the scale of v/v_o.*

The evolution of atmospheres

negligible, and even if all of the Earth's oxygen could reach the critical escape level, it would not escape the pull of gravity in the age of the planetary system. For practical purposes, therefore, we can neglect the leakage and consider oxygen to be a permanent constituent of the Earth's atmosphere.

The situation is different for hydrogen, however. Hydrogen is present in the Earth's atmosphere almost exclusively in the form of water. Any large amount of free hydrogen that may once have existed has escaped during the history of the planet. For hydrogen, v_0 is four times as great as for oxygen. This difference is enough to increase the proportion of atoms whose velocities exceed v_e by a factor of 10^{79}, thereby making hydrogen an ephemeral species, at least on a cosmic time scale.

Now we see how Venus may have lost its water. The runaway greenhouse effect could have driven the water into the atmosphere where it would have been subject to photodissociation by solar ultraviolet radiation. The hydrogen produced could then have escaped to space, leaving behind oxygen, which could have been consumed in reactions with surface minerals. This explanation is possible but we cannot, of course, be sure that it is correct.

Whereas hydrogen is an ephemeral species on the inner planets, conditions are very different on the outer planets. For these planets low temperatures (see Table 3–1) give low values for the most probable velocity v_0, and large surface gravities (see Table 1–5) give large values for the escape velocity v_e. The result is a negligibly small fraction of hydrogen atoms able to escape. The outer planets evidently formed mainly from hydrogen, which was the most abundant constituent of the primordial solar system, and this hydrogen has remained dominant in their atmospheres to the present day (see Table 1–2).

On the other hand, because of their small gravities, the satellites of the inner planets should retain no atmospheres at all. In the case of Mercury, the gravitational acceleration is small and the temperature is high; we do not anticipate any atmosphere, but discriminating measurements have not yet been made.

Terrestrial Life and Atmospheric Oxygen

Living organisms on Earth are another factor having a great effect on atmospheric composition. As an example of the interaction between the atmosphere and the biosphere we shall discuss the cycles of atmospheric oxygen (see Fig. 6–5).

When water is photodissociated and the hydrogen escapes to space, oxygen is left behind in the atmosphere. This source is shown at the top of

The evolution of atmospheres

Fig. 6–5. It is unimportant compared with the source provided by photosynthesis (the process discussed earlier in this chapter in connection with the cycles of carbon dioxide).

Because of the dominant role of photosynthesis as a source of oxygen, there is a close connection between the oxygen cycles and the carbon cycles. Photosynthesis releases oxygen to the atmosphere and removes carbon dioxide. Respiration and decay remove oxygen from the atmosphere and release carbon dioxide. If life were to cease on Earth and all living things were to decay, the carbon they contain would consume about 10^{19} gm of atmospheric oxygen.

Most organisms do eventually decay, consuming just as much atmospheric oxygen as was released to the atmosphere by photosynthesis when they were growing. Only the small fraction of organic debris that is buried in sediments without decaying corresponds to an overall addition of oxygen to the atmosphere. The oxygen source corresponding to the burial of organic carbon is indicated in the bottom right corner of Fig. 6–5. Oxygen, of course, does not go on accumulating in the atmosphere indefinitely. Ultimately the organic carbon is brought back to the surface as mountains

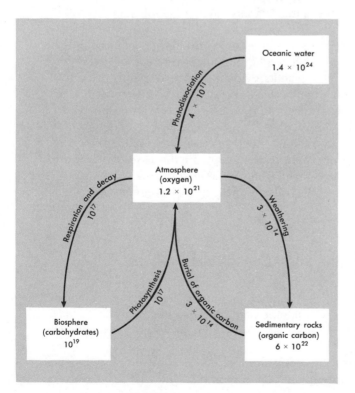

FIGURE 6-5 *The cycles of oxygen. Boxes denote reservoirs of oxygen; the contents of the reservoirs are expressed in gm of O_2. The arrows denote the transfer of material between reservoirs; the rates of transfer are given in gm of O_2 per year.*

The evolution of atmospheres

are raised up and eroded. Once the sedimentary rocks are exposed to the atmosphere, the organic carbon they contain reacts with oxygen to produce carbon dioxide, thus closing the cycle of burial and weathering.

The interesting difference between oxygen and carbon dioxide lies in the stability of the amount in the atmosphere. We have already described how the amount of carbon dioxide in the atmosphere is controlled by the ocean, and we have pointed out the stabilizing effect of the enormous oceanic reservoir. A large amount of carbon dioxide can be released to the atmosphere without causing long-term changes because the extra gas dissolves in the ocean, but there is no equivalent stabilizing reservoir for oxygen. The oxygen content of the ocean is relatively small because oxygen is much less soluble than carbon dioxide. It is possible, therefore, that there have been marked fluctuations in the level of atmospheric oxygen in the course of geologic history.

In particular, in the time before life evolved on Earth there can have been very little oxygen in the atmosphere at all. The small amount produced as a result of photodissociation of water vapor and escape of hydrogen must have been consumed almost at once by the weathering of rocks. Without oxygen there can have been no protective screen of ozone in the atmosphere, and solar ultraviolet radiation in the 2000 Å to 3000 Å wavelength range may have penetrated to the ground. Since this radiation is lethal to simple organisms, conditions were most inhospitable.

In spite of this, primitive plants did develop and start to photosynthesize. Their remains have been found in rocks that are at least two billion years old. In time the plants were able to produce more oxygen by photosynthesis than was consumed by rock weathering, and oxygen began to accumulate in the atmosphere. We do not know when this happened, but it must have preceded the flowering of life that marks the opening of the Paleozoic era, about 600 million years ago.

We still have much to learn about the history of the Earth's atmosphere. The ideas we have described are speculative, but they are the first steps in a process that will enhance our understanding of the factors that have made life on Earth possible.

Appendix

Temperature of the Stratopause

We have remarked on a number of occasions that the high temperature near 50 km is caused by absorption of solar radiation by ozone. Let us examine this statement a little further in terms of Fig. 3–9, which refers to the transparent outer fringe of the atmosphere.

We drop the assumption that solar radiation passes directly to the Earth's surface without absorption. Instead, we allow the thin layer of atmosphere to absorb visible solar radiation with an opacity δ. For infrared planetary radiation (see Fig. 3–4), the opacity of the layer is ε. The solar flux can be equated to σT_e^4, so that $\delta \sigma T_e^4$ is the amount of solar energy absorbed in the layer.

The balance of energy is now altered. We must write

Energy absorbed from solar radiation + Energy absorbed from infrared planetary radiation = Energy lost by infrared radiation

or

$$\delta \sigma T_e^4 + \varepsilon \sigma T_e^4 = 2\varepsilon \sigma T_s^4$$

therefore

$$\left(\frac{T_s}{T_e} \right)^4 = \frac{\delta + \varepsilon}{2\varepsilon}$$

Our previous discussion corresponded to $\delta = 0$ (no absorption of solar radiation in the atmosphere) so the skin temperature was given by $(T_s/T_e)^4 = \frac{1}{2}$. The optical properties of the atmosphere for solar and planetary radiation, however, are quite different. In particular, the opacity for solar radiation can be very great indeed at some wavelengths. Thus, δ can be greater than ε at high altitudes, leading to higher temperatures at the altitudes where solar radiation is absorbed.

In this way we can understand the existence of warm layers in the atmosphere. Figure 2–4 tells us that the warm layer caused by ozone absorption should occur close to 50 km above the ground. Detailed calculations of the temperature profile give good agreement between theory and observation.

The Adiabatic Lapse Rate

A change in which no heat transfer takes place is called *adiabatic*. We wish to discover the rate of variation of temperature with height consistent with adiabatic motions; we have already demonstrated that the temperature will decrease with height.

In Fig. A–1 we wish to move a parcel of air adiabatically from height z_1 (pressure p_1, temperature T_1) to height z_2 (pressure p_2, temperature T_2), without exchanging any heat with the surrounding atmosphere. We wish to find the value of the adiabatic lapse rate,

$$\Gamma = -\left(\frac{T_2 - T_1}{z_2 - z_1}\right)$$

Consider a parcel consisting of a single gram of air. When it undergoes any change of height, temperature, or pressure, the first law of thermodynamics states that the total energy in all identifiable forms (erg gm^{-1}) must not change. Let us consider what these changes are:

1. Q = heat brought into the parcel. For an adiabatic change, $Q = 0$.

2. E = increase of internal energy of the molecules. It is a property of a perfect gas that internal energy depends on the temperature only. Thus, for a change at constant temperature, $E = 0$.

3. W = work done on the parcel. Since the gas is compressible, the boundaries of the parcel will move inward if pressure is applied. The parcel will contract and work will be done on it. For an incompressible substance, however, $W = 0$.

4. P = increase of potential energy in the gravitational field. This is the work that could be realized in falling from z_2 to z_1. The force on 1 gm (dyne gm^{-1}) is $g \times 1$ (cm sec^{-2}). Work is force times distance. Thus, the potential energy of a gram at z_1 is less than that at z_2 by the amount $g \times (z_2 - z_1)$ (cm^2 sec^{-2} equal to erg gm^{-1}).

The first law of thermodynamics, in symbolic form, states that for any change

$$P + E = Q + W$$

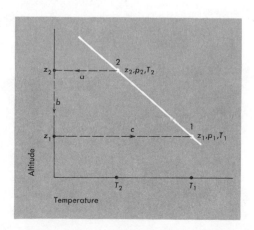

FIGURE A-1. *Evaluation of the adiabatic lapse rate.*

If we pass 1 gm of air directly from 2 to 1, we must have $Q = 0$, since the change is adiabatic. To evaluate the terms P, E, and W, we will go from 2 to 1 by changes a, b, and c. Changes a and c take place at constant height, or constant pressure (the two are related through the barometric law). On the other hand, b is a change of height at the absolute zero of temperature.

Consider changes a and c. The amount of heat required to change the temperature of 1 gm of air by 1°K is called the *specific heat* (erg gm^{-1} deg^{-1}). If the change takes place at constant pressure, this specific heat is known as c_p, the specific heat at constant pressure. It is an important property of an ideal gas that c_p is a constant, and does not depend upon temperature or pressure. This is, in effect, an evaluation of the two terms $(W - E)$. The term P we know to be zero if the height does not change. Hence, for a,

$$Q_a = - c_p T_2 \text{ (heat is removed)}$$

and for c

$$Q_c = +c_p T_1 \text{ (heat is added)}$$

For step b, $E = 0$ because the change is at constant temperature. Now, as we are aware from common experience, at very low temperatures a gas will assume the high density of a solid. In fact, if the ideal gas law were obeyed at all temperatures, the density would become infinitely large at $T = 0$. The important point is that when the parcel is small and very dense it is effectively incompressible, like a solid or a liquid. Hence for change b, $W = 0$, and

$$Q_b = P = - g(z_2 - z_1)$$

If the three changes together are to be adiabatic, we must have

$$Q_a + Q_b + Q_c = 0$$

Therefore

$$c_p (T_1 - T_2) - g(z_2 - z_1) = 0$$

and

$$\Gamma = - \left(\frac{T_2 - T_1}{z_2 - z_1}\right) = g/c_p$$

Suggestions for further reading

Barry, R. G., and R. J. Chorley, 1970, *Atmosphere, Weather, and Climate.* New York: Holt, Rinehart & Winston.

Bates, D. R., 1964, *The Planet Earth.* New York: Pergamon Press.

Battan, L. J., 1961, *The Nature of Violent Storms.* Garden City: Doubleday.

Battan, L. J., 1962, *Cloud Physics and Cloud Seeding.* Garden City: Doubleday.

Battan, L. J., 1962, *Radar Observes the Weather.* Garden City: Doubleday.

Battan, L. J., 1966, *The Unclean Sky.* Garden City: Doubleday.

Battan, L. J., 1969, *Harvesting the Clouds: Advances in Weather Modification.* Garden City: Doubleday.

Craig, R. A., 1968, *The Edge of Space.* Garden City: Doubleday,

Day, J. A., 1970, *The Science of Weather.* Reading, Mass.: Addison-Wesley.

Day, J. A., and G. L. Sternes, 1970, *Climate and Weather.* Reading, Mass.: Addison-Wesley.

Eshleman, Von R., "The Atmospheres of Mars and Venus," *Scientific American,* Vol. 220, No. 3 (March 1969), 79–88.

Fairbridge, R. W., ed., 1967, *The Encyclopedia of Atmospheric Sciences and Astrogeology.* New York: Reinhold.

Hidy, G. M., 1967, *The Winds.* Princeton: Van Nostrand.

Longstreet, T. M., 1962, *Understanding the Weather.* New York: Collier Books.

Lorenz, E. N., "The Circulation of the Atmosphere," *American Scientist,* Vol. 54, No. 4 (December 1966), 402–420.

Ludlam, F. H., and R. S. Scorer, 1957, *Cloud Study.* London: John Murray.

Mason, B. J., 1962, *Clouds, Rain and Rainmaking.* London: Cambridge University Press.

Mehlin, T. G., 1968, *Astronomy and the Origin of the Earth.* Dubuque, Iowa: Wm. C. Brown.

Miller, A., 1971, *Meteorology.* Columbus, Ohio: Charles E. Merrill.

Newell, R. E., "The Circulation of the Upper Atmosphere," *Scientific American,* Vol. 210, No. 3 (March 1964), 62–74.

Ohring, G., 1966, *Weather on the Planets.* Garden City: Doubleday.

Peek, B. M., 1958, *The Planet Jupiter.* London: Faber & Faber.

Ratcliffe, J. A., 1970, *Sun, Earth and Radio.* New York: McGraw-Hill.

Rumney, G. R., 1970, *The Geosystem: Dynamic Integration of Land, Sea and Air.* Dubuque, Iowa: Wm. C. Brown.

Shapiro, I. I., "Radar Observations of the Planets," *Scientific American,* Vol. 219, No. 1 (July 1968), 28–37.

Slipher, E. C., 1962, *The Photographic Story of Mars.* Cambridge, Mass.: Sky Publishing.

Sutcliffe, R. C., 1966, *Weather and Climate.* New York: W. W. Norton.

Sutton, O. G., 1962, *The Challenge of the Atmosphere.* London: Harper.

Index

Mars   65-67
       98-99
       133-34